TEACH YOURSELF BOOKS

Bridge

Bridge

Terence Reese

Revised by David Bird

TEACH YOURSELF BOOKS

For UK order queries: please contact Bookpoint Ltd, 39 Milton Park, Abingdon, Oxon OX14 4TD. Telephone: (44) 01235 400414, Fax: (44) 01235 400454. Lines are open from 9.00–6.00, Monday to Saturday, with a 24 hour message answering service. Email address: orders@bookpoint.co.uk

For U.S.A. & Canada order queries: please contact NTC/Contemporary Publishing, 4255 West Touhy Avenue, Lincolnwood, Illinois 60646–1975 U.S.A. Telephone: (847) 679 5500, Fax: (847) 679 2494.

Long-renowned as the authoritative source for self-guided learning – with more than 30 million copies sold worldwide – the *Teach Yourself* series includes over 200 titles in the fields of languages, crafts, hobbies, sports, and other leisure activities.

A catalogue record for this title is available from The British Library.

Library of Congress Catalog Card Number: On file

First published in UK 1980 by Hodder Headline Plc, 338 Euston Road, London NW1 3BH

First published in US 1992 by NTC/Contemporary Publishing, 4255 West Touhy Avenue, Lincolnwood (Chicago), Illinois 60646–1975 U.S.A.

The 'Teach Yourself' name and logo are registered trade marks of Hodder & Stoughton Ltd.

Cover photo © Barbara Baran

Typeset by Transet Limited, Coventry, England.
Printed in Great Britain for Hodder & Stoughton Educational, a division of Hodder Headline Plc, 338 Euston Road, London NW1 3BH by Cox & Wyman Ltd, Reading, Berkshire.

Impression number	10	9	8	7	6	5	4	3		
Year		2004		2003	2002	2001	2000		1999	

CONTENTS

FOREWORD

Bridge is one of the world's finest card games, a source of endless fascination, and an entry into a world of new friends.

How good at the game will you be after absorbing the contents of this book? The game consists of two parts – the bidding and the play of the cards. Bidding is easily learnt from a book. By the time you have turned the final page, your bidding will be better than that of the majority of bridge players. Learning to play the cards well is more difficult and takes time. We cover the basics here. You will find that your cardplay improves gradually, every time that you play.

One thing is certain. You will never regret the day that you first took up bridge, the king of card games.

Terence Reese
David Bird

1 | IF THE GAME IS NEW TO YOU

Bridge is the most famous card game in the world, rivalled for the title only by Poker. It is a game for four players. The players who sit opposite each other are partners and compete against the other two players.

Thirteen cards are dealt face down to each player and the first task is to 'sort out' the hand. Each player picks up his cards and places together all the cards in the same suit. The rank of the cards is the standard one:

(highest) Ace King Queen Jack 10 9 8 7 6 5 4 3 2 (lowest)

Within a suit it is normal to sort your cards from left to right in descending order of rank, so you might sort out your hand to look like this:

♠ A Q 6 2   ♥ 7 5 2   ♦ K Q 10 2   ♣ K 4

A bridge player, describing this hand, would say 'I held four spades to the ace-queen, three small hearts, four diamonds to the king-queen-10 and king doubleton of clubs.'

Once all four players have sorted their hands, the 'bidding' takes place. We will see in a moment how players make their bids, but the general purpose is for a partnership to determine how many tricks they will make when the cards are played, also whether one of the four suits should be made trumps. (Don't worry if the notion of 'tricks' or 'trumps' is new to you. We will cover this in a moment.)

Once the bidding is over, it is time for the cards to be played. The partnership who made the highest bid during the bidding try to make the number of tricks that they said they could. The score is calculated and then another hand is dealt.

That was a brief summary of a hand of bridge. It will be easier to understand the bidding if we describe first how the play proceeds. Should you already be familiar with the idea of tricks and trumps, you may assume a superior expression and skip to page 4.

What is a 'trick'?

The four players sitting at the card table are referred to as North, East, South and West in bridge books and newspaper columns. A trick consists of one card from each of the four players. The cards are played in turn, around the table, in a clockwise direction. If West is first to play to a trick, North will play the second card, East the third card, and South the last card.

This is a typical trick, with West playing first (or leading to the trick, as bridge players say):

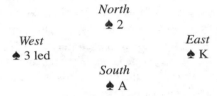

West leads the 3 of spades. North plays the 2, East the king, and South the ace. South 'wins the trick' because he played the highest card. The four cards are gathered together and placed face down on the pile of tricks for North–South. (North and South are acting as a partnership, remember.)

As you saw, everyone played the same suit, here a spade. It is a requirement of the game that you must play the same suit as that of the card led. If a club is led, for example, you must play a club if you have any clubs in your hand.

Each player starts with 13 cards. So, when all the cards have been played 13 tricks will have been won between the two sides. If North–South held roughly the same number of high cards (such as aces and kings) as East–West, they might perhaps score seven tricks and East–West would score the remaining six. If instead North–South were dealt nearly all the high cards they might score 12 tricks and the other side only one. When we come to look at the bidding, we will see that the two sides have to estimate how many tricks they will take. They have to do this before the play actually starts.

When you cannot follow suit

Suppose you are dealt only two spades. On the first two occasions when a spade is led, you will be able to follow with a spade. The third time a spade is led, you will have no spade to play. You are then allowed to play any card you wish from a different suit.

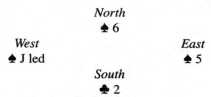

We will assume that two rounds (tricks) of spades have already been played. On the third round West leads the jack, which is the highest spade outstanding. North and East follow suit with smaller cards. South has no spades left in his hand, so he 'discards' or 'throws' a club.

West wins the trick because he played the highest spade on the trick. He would have won the trick even if South had discarded the ace of clubs. In the absence of a trump suit (which will be explained in the next section) a trick is won by the highest card played in the suit that was led. So, it is possible to win a trick with a 2! If no one can follow suit, you can lead a 2 and score a trick with it.

Whoever wins one trick is first to play to the next. Suppose this is the first trick of a hand:

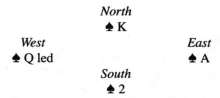

West leads the queen of spades. North 'covers' with the king and East wins the trick with the ace, South following with a small card. Since East won this trick he will be first to play to the next trick. It is entirely his choice whether he leads a spade to the next trick, or some other suit.

What are 'trumps'?

On most hands, but not all, one of the four suits becomes trumps. The trump suit is more powerful than all the other suits. Any trump, even the 2, has the power to beat any card in a different suit. Players must always follow suit if they can, but when they have no card remaining in the suit led, they can play a trump. This will win the trick, unless someone subsequently plays a higher trump to the trick.

Let's suppose that hearts are trumps and this is the first trick of the hand:

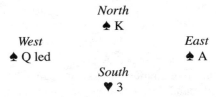

West leads the queen of spades, covered by the king and ace. South was dealt no spades (he was 'void' in spades, as bridge players say). He is therefore allowed to trump, or 'ruff', the trick. Hearts are trumps and South's ♥ 3 wins the trick.

This is what may happen when there are two players who cannot follow suit (again hearts are trumps):

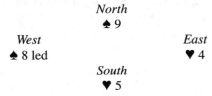

West leads ♠ 8 and this is beaten by North's ♠ 9. East has no spades left and chooses to ruff with the 4. South also has no spades left. He 'overruffs' with the 5 and wins the trick.

The play of a whole hand in no-trumps

No doubt you are itching to hear about the bidding. Not long to wait! For the moment we are going to look at the play of two whole hands.

The first will be played in no-trumps (without any suit as trumps). As a result of the bidding, South must try to make nine tricks. He is known as the 'declarer' and will play both his own cards and those of the hand opposite (known as 'the dummy'). West will lead to the first trick.

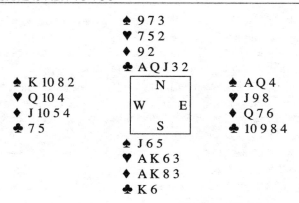

♠ 9 7 3
♥ 7 5 2
♦ 9 2
♣ A Q J 3 2

♠ K 10 8 2   ♠ A Q 4
♥ Q 10 4   ♥ J 9 8
♦ J 10 5 4   ♦ Q 7 6
♣ 7 5   ♣ 10 9 8 4

♠ J 6 5
♥ A K 6 3
♦ A K 8 3
♣ K 6

Trick 1 In no-trumps it is normal to lead from your strongest suit. West leads ♠ 2 and North now lays out his hand, which will become the dummy. He arranges the cards not as you see them in this diagram, but like this:

♠	♥	♦	♣
9	7	9	A
7	5	2	Q
3	2		J
			3
			2

Declarer reaches forward to play a low spade from dummy. East plays the ace of spades, winning the trick, and the declarer follows with the 5.

Trick 2 Since East won the first trick, he leads to the second. He leads the queen of spades and the other three players follow with the 6, 8 and 7.

Trick 3 East is still on lead. He leads his last spade and West captures South's jack with the king.

Trick 4 West leads ♠ 10. This is the last spade and since no suit has been made trumps on this hand, it must win the trick. Declarer plans to make tricks with dummy's club suit later, so he throws a heart (or a diamond) from the dummy. The next two players throw a heart too.

Trick 5 Declarer's target (set in the bidding) was nine tricks. If the defenders could score one more trick, they would defeat him. Unfortunately for West, he does not have another trick to take. He leads ♦ J, South winning with the king.

Trick 6 Declarer now seeks to make five tricks in clubs. To achieve this he must take one club trick with the king, then four more with dummy's clubs. (It usually works best if you play first the high cards in the shorter holding.) South leads ♣ K, the other three hands following with low clubs.

Trick 7 Declarer leads ♣ 6 and wins the trick with dummy's ace. You see the advantage of playing the king on the first round? The lead is now in dummy and declarer will be able to make tricks with the other high clubs.

Tricks 8 and 9 Declarer wins the next two tricks by leading the queen of clubs, and then the jack of clubs.

Trick 10 Declarer leads dummy's last club, the 3. No one else has any clubs left, so this lowly card wins the trick, everyone else discarding.

Trick 11 Declarer leads a heart from dummy and wins the trick with the ace.

Trick 12 Declarer wins the penultimate trick with the king of hearts.

Trick 13 Declarer wins the last trick with the ace of diamonds.

What was the effect of all that? The defenders scored four tricks and the declarer scored nine. Since that was his target, he was successful on this occasion. Bridge players would say, 'South scored nine tricks in no-trumps.'

When you are new to the game, it's not easy to follow the play of a hand from a diagram. You will soon get used to it. Look back to the full diagram. If you showed it to a bridge player and asked him how many tricks could be made in no-trumps, he would say, 'You can make nine tricks: five in clubs, and four more from the ace-kings in the red suits.'

The play of a whole hand, with trumps

The presence of a trump suit has a big effect on the play. On the next deal North–South have ten spades between them. In the bidding (yes, we know we haven't explained that yet!) they have decided to make the spade suit trumps. They have also set themselves a target of ten tricks. Let's see if South, the declarer, can manage it.

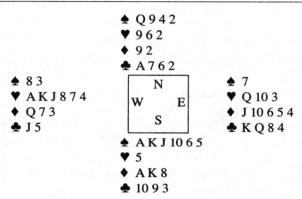

♠ Q 9 4 2
♥ 9 6 2
♦ 9 2
♣ A 7 6 2

♠ 8 3
♥ A K J 8 7 4
♦ Q 7 3
♣ J 5

♠ 7
♥ Q 10 3
♦ J 10 6 5 4
♣ K Q 8 4

♠ A K J 10 6 5
♥ 5
♦ A K 8
♣ 10 9 3

Trick 1 West leads ♥ A, which wins the first trick.

Trick 2 West leads ♥ K. This does not win the trick. South has no hearts left and is therefore allowed to trump (or 'ruff', as most players say). He ruffs with ♠ 5, winning the trick. This is one of the big advantages of having a trump suit; you can stop the opponents from making tricks in their own strong suit.

Trick 3 Declarer's next move is to remove the defenders' trumps. He leads the ace of trumps, everyone following.

Trick 4 Declarer 'draws a second round of trumps'. He leads a low trump from his hand and wins with dummy's queen of trumps, East discarding a diamond.

Trick 5 Declarer makes a trick by leading a low diamond to his ace.

Trick 6 Declarer makes a trick by leading ♦ K.

Trick 7 Declarer leads ♦ 8, West producing the queen. Now we see the other main advantage of a trump suit. Declarer can score an extra trick by ruffing in the dummy. He plays dummy's ♠ 4, winning the trick.

Trick 8 Declarer makes a trick by leading ♣ A.

Trick 9 Declarer has made seven tricks already and will eventually make three more with the remaining three trumps in his hand. He leads ♣ 2 and East wins the trick with the king.

Trick 10 East makes a trick by leading ♣ Q.

Trick 11 East leads ♥ Q, declarer ruffing with the 10.

Trick 12 Declarer makes a trick by leading the king of trumps.

Trick 13 Declarer makes a trick by leading the jack of trumps.

The end effect this time? Declarer made ten tricks with spades as trumps. Once again he was successful in achieving his target.

If we showed the full diagram to a bridge player asking how South would fare with spades as trumps, the reply would be: 'It looks like ten tricks. Declarer can make six trump tricks in the South hand, three winners in clubs and diamonds, and one diamond ruff in the dummy.'

The great moment has arrived. Now that you have some idea of how a hand is played, both in no-trumps and with one suit as trumps, it is time to take our first look at the bidding.

The purpose of the bidding

Think back to the two hands we have just seen played. On one, North–South played in no-trumps and scored nine tricks. On the other, they played with spades as trumps and scored ten tricks. The purpose of a partnership's bidding is two-fold: to decide which suit should be made trumps, if any, and to estimate how many tricks they will be able to make.

A 'bid' is a way of describing your hand to your partner. By telling each other which suits you hold length in, you hope to find a good trump suit. By telling each other how strong you are (in other words, whether you have a lot of high cards), you hope to be able to estimate how many tricks you can make. All of this is accomplished without seeing each other's hands, of course. You exchange information only by making bids.

So, how do you make a bid? A bid is a combination of a number and a suggestion of what should be trumps. 'One Heart' is a bid, so is 'Three No-trumps'.

As well as describing your hand, a bid is an undertaking to score a number of tricks with the named suit as trumps. There are 13 tricks to be won in all and the lowest number you are allowed to attempt is seven. Since the first six tricks are assumed, a bid One Heart means that you think you can make seven tricks with hearts as trumps. Two Spades means you are willing to try for eight tricks with spades as trumps. Three No-trumps? Yes, nine tricks without any suit as trumps.

The number of tricks implied by a bid is always six more than the number included in the bid. If you made a bid at the seven level, such as Seven Clubs, you would have to score all 13 tricks.

When can I open the bidding?

To measure how strong a hand is, in terms of high cards, some sort of scale is needed. This valuation is used worldwide:

    Ace   – 4 points
    King   – 3 points
    Queen   – 2 points
    Jack   – 1 point

There are ten points in each suit, 40 in the whole pack. Since the cards are shared equally between the four players, an average hand will contain ten points. A sound general rule is that you should open the bidding when you hold 12 points or more.

You will soon become familiar with counting the points in your hand. How many points does this hand contain?

                    ♠ A Q 10 2
                    ♥ K 4
                    ♦ J 3
                    ♣ A K 8 3 2

There are 17 high-card points. Your best suit is clubs and you would open 'One Club'. What about this hand?

                    ♠ A K 10 9 6 3
                    ♥ 6 2
                    ♦ K J 4 3
                    ♣ 5

You have 11 points, normally not quite enough to open the bidding. Here, though, you have a very good six-card spade suit. Long suits help you to make tricks in the play and increase the value of your hand. You would open 'One Spade'.

Suppose instead that you were dealt this moderate collection:

                    ♠ 8 2
                    ♥ Q 9 8 6 2
                    ♦ 10 4
                    ♣ K J 6 2

It is a poor hand with fewer high cards than average. You would say 'No bid' (or 'Pass' in USA) and your partner would then know that your hand is relatively weak.

How the auction proceeds

So far we have looked only at the opening bid, the first bid in the auction. As you will have guessed from the word 'auction', there can be many bids on one deal. The dealer has the first opportunity to make a bid, then the player to his left, and so on. The bidding goes round in a circle, clockwise. As with most auctions your bid, if any, must be higher than the previous bid. The bidding continues, up and up, until there are three passes in succession.

You may wonder if One Spade is a higher bid than One Diamond. It is, in fact. This is the ranking order of the five denominations:

No-trumps	(highest)
Spades	
Hearts	
Diamonds	
Clubs	(lowest)

So, if the previous bid was One Club, you can bid One of any other suit, or One No-trump. But if the previous bid was One Spade and you want to bid clubs, you would have to bid at least Two Clubs.

It often happens that both partnerships want to enter the bidding. Here is a full deal with a typical competitive auction:

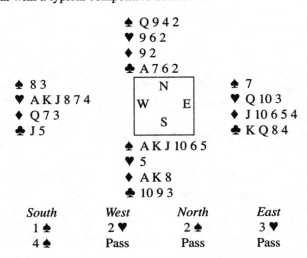

```
                        ♠ Q 9 4 2
                        ♥ 9 6 2
                        ♦ 9 2
                        ♣ A 7 6 2
        ♠ 8 3               N            ♠ 7
        ♥ A K J 8 7 4                    ♥ Q 10 3
        ♦ Q 7 3        W        E        ♦ J 10 6 5 4
        ♣ J 5              S             ♣ K Q 8 4
                        ♠ A K J 10 6 5
                        ♥ 5
                        ♦ A K 8
                        ♣ 10 9 3
```

South	*West*	*North*	*East*
1 ♠	2 ♥	2 ♠	3 ♥
4 ♠	Pass	Pass	Pass

South has 15 points and a good spade suit. He opens the bidding with One Spade. West has a good heart suit and enough strength to enter the bidding. Since hearts rank below spades he has to bid Two Hearts. North has a spade fit with his partner; he 'supports' his partner by bidding Two Spades. Similarly, East supports his partner.

Each bid is higher than the preceding one, as you see. Once his partner has raised him to Two Spades, South thinks he will be able to make ten tricks with spades as trumps. He therefore bids Four Spades. No one wants to bid any more. Three consecutive passes end the auction.

Four Spades is the final 'contract'. The player who made the first bid in the trump suit, South on this occasion, will become the declarer. The player on his left, West, will make the opening lead.

Do you recognise the deal? It was one we looked at earlier. If you look back to page 7 you will see that South is going to 'make his contract'. He will indeed score ten tricks in spades.

Making a 'game'

You may wonder why, on the previous hand, South set himself a target of ten tricks rather than just nine. After all, the higher target you set yourself, the more chance there is that you will fail. The answer lies in the scoring system used for bridge.

Each trick that you bid and make, beyond the first six, has a scoring value. It varies, depending on what was trumps. Spades and hearts are known as the 'major suits'; diamonds and clubs are the 'minor suits'. Tricks are worth more when a major suit is trumps. This is the scale:

> Diamonds or clubs are trumps: each trick scores 20
> Spades or hearts are trumps: each trick scores 30
> No-trumps: 1st trick scores 40, each other trick 30

The prime objective at bridge is to 'make a game'. To do this you must score 100 points. The important point to note is that tricks only count towards game if you said you would make them during the bidding.

Suppose you play with spades as trumps. To make a game you would have to make Four Spades (ten tricks), because $4 \times 30 = 120$. And it's not enough simply to score ten tricks, you must also bid Four Spades. If, say, you were to bid Three Spades and actually make ten tricks you would only score $3 \times 30 = 90$ points towards game.

Think of the level of a contract as setting the bar in high jump. If you set the bar high and can clear it, you will get a big score. If you knock the bar down (by not scoring as many tricks as you said you would), you score nothing.

If the contract is set below the game level – Two Diamonds, for example – then the points you score are left on the scoresheet and will count towards the 100 you need for game. The contract is called a 'part score'. A part score of 40, followed by another of 60, would give you a game.

We will not bother ourselves further with the scoring at this stage. We will just say that you get a big bonus for making a game. An even bigger bonus is awarded if you bid to the six level and make it, known as a 'small slam'. The biggest bonus of all is awarded if you bid to the seven level and make it, a 'grand slam'.

These are the possible final contracts:

Part scores:	1♣	1♦	1♥	1♠	1NT
	2♣	2♦	2♥	2♠	2NT
	3♣	3♦	3♥	3♠	
	4♣	4♦			
Games:					3NT
			4♥	4♠	4NT
	5♣	5♦	5♥	5♠	5NT
Small slams:	6♣	6♦	6♥	6♠	6NT
Grand slams:	7♣	7♦	7♥	7♠	7NT

The easiest route to game is to make 3NT, to score nine tricks without any suit as trumps. Next easiest is 4♥ or 4♠, where you need ten tricks. With clubs or diamonds as trumps you need to bid 5♣ or 5♦. This is not a very attractive game to bid, since you will need to score 11 tricks out of 13.

Points to remember

1 A bid consists of a number and a suggested choice of trumps. Three Spades, for example, means that you think you can make nine (6 + 3) tricks with spades as trumps.

2 Each bid must be higher than the preceding one. The bidding ends when there are three consecutive passes.

3 The declarer is the player who made the first bid in the chosen denomination (the trump suit, or no-trumps). He must attempt to make the 'final contract'. The player to declarer's left will make the opening lead. The next player's hand, the dummy, will then be placed face up on the table.

4 The primary objective in bridge is to make a 'game', to score 100 points. 3NT is game contract; so are 4♥, 4♠, 5♣ and 5♦. To score 100 points you have also to bid the game. Only tricks that were bid will count towards the 100 points for game.

5 A contract worth less than 100 points is known as a 'part score'. The points gained still count towards a game. If you make a part score of 60, a subsequent part score of 40 will give you a game.

Quiz

a Of the four suits, which ranks second highest?

b Suppose someone bids 'Three Diamonds'. How many tricks would they have to make if this became the final contract?

c How many points do you need to make game?

d If spades are trumps, how many tricks do you need for game?

e How many high-card points are there in this hand:

   ♠ A Q J 5 3
   ♥ 10 8 3
   ♦ 7
   ♣ A K 6 2

f Would you open the bidding on the hand in **e**? If so, what bid would you make?

Answers

a Hearts.

b Nine.

c 100.

d Ten.

e 14.

f Yes. You would open One Spade.

2 | MAKING TRICKS IN A SINGLE SUIT

Before saying anything more about the bidding, we will look at some of the ways in which you can score extra tricks in the play. Until you have a rough idea of how tricks are won and plans are formed, it will be hard to understand why, for example, it is reasonable to bid to 3NT on two particular hands in combination.

You need to be able to recognise how many tricks you can expect from a particular suit. Sometimes it is easy to work out:

North
♠ A Q 6

South
♠ K 8 3

Clearly you can make three spade tricks, one with each high card. But that is only the beginning. There are numerous ways of establishing tricks from less substantial holdings.

Establishing a suit

When the opponents hold a high card in the suit you want to establish, you will have to 'knock it out'. Here the high card you are missing is the ace:

North
♥ Q J 7

South
♥ K 8 3

So long as you do not commit the folly of playing two high cards on the same trick, you can be sure of two tricks. You can play a low card to the queen. If the ace does not appear, you lead the jack to the next trick. The defenders will score one trick in the suit, you will score two. You 'knocked out the ace' and 'established two heart tricks'.

The play is much the same here:

North
♦ Q J 10 5

South
♦ K 4

By leading the king on the first round, then the 4, you can establish three winners. As mentioned before, it is generally best to play first any high cards in the shorter holding. Suppose instead that you started with the queen and a defender won with the ace. Your king would win the next trick and the lead would be in the wrong hand.

You may need to knock out more than one high card.

North
♣ Q 10 8 5

South
♣ J 9 2

After the defenders have taken their ace and king, you will win two tricks.

Here you are missing three high cards:

North
♠ J 9 7 4 3

South
♠ 10 8

Given time, you will be able to establish two winners, by knocking out the ace, king and queen.

The number of tricks that you can establish will often depend on how the opponents' cards 'break' (or divide).

North
♣ 9 5 4 3

West   *East*
♣ Q J   ♣ A K

South
♣ 10 8 7 6 2

Here the defenders hold four honours. Because of the 2–2 break (each defender holds two cards) they will make only two tricks. You lead one round of the suit and East wins with the king. When you regain the lead and play a second round, East wins with the ace and the defenders then have no clubs left. You will score three tricks with the remaining small cards. They will be 'good', as bridge players say.

Had the defenders' cards broken 3–1, the situation would be less favourable:

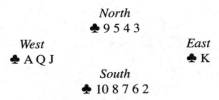

North
♣ 9 5 4 3

West
♣ A Q J

East
♣ K

South
♣ 10 8 7 6 2

Now you would have to play three rounds of the suit to knock out the defenders' honours. You would then establish two tricks for your small cards.

Sometimes you need to surrender one or more tricks to the opponents, even when you hold the top cards yourself.

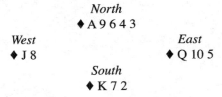

North
♦ A 9 6 4 3

West
♦ J 8

East
♦ Q 10 5

South
♦ K 7 2

You can make two tricks with the king and ace, then let East make a trick with the queen. By this time, the defenders will have no cards left in the suit. The last two cards in dummy will be good and you will make a total of four tricks from the suit.

The defenders' cards may not be so helpfully divided:

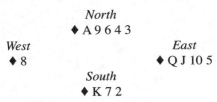

North
♦ A 9 6 4 3

West
♦ 8

East
♦ Q J 10 5

South
♦ K 7 2

After making tricks with the king and ace, you would have to let East win two tricks in order to establish a third diamond trick for yourself.

Leading towards high cards

The chance of developing extra tricks will often depend on the position of the defenders' cards. The general principle is to lead *towards* high cards, not away from them. This is the simplest example:

North
♥ K 5

South
♥ 6 4 3

You should lead a low card towards the king in the North hand. If the ace is held by West, the king will win a trick. If instead you led the suit from dummy, you would never make a trick with the king.

North
♠ Q 6 3

South
♠ A 7 2

To make two tricks from this combination you must lead towards the queen, hoping that West holds the king. There would be no point at all in leading the queen from dummy. East would cover with the king, if he held it; you would make only one trick wherever the king was.

The idea is the same when you are missing two high cards.

North
♦ Q J 6

West   *East*
♦ A 5 2   ♦ K 10 8 3

South
♦ 9 7 4

To establish a trick you must lead twice towards the high cards in dummy. On the first round East will win the queen with the king. When you subsequently lead towards the jack, you cannot be stopped from making one trick from the suit.

It is often possible to combine the techniques of leading towards high cards and establishing low ones:

North
♣ K Q 7 5 2

West   *East*
♣ A 10 4   ♣ J 9

South
♣ 8 6 3

You lead towards dummy's club honours. West plays low and dummy's king wins. You return to the South hand, by playing some other suit, and lead a second round of clubs. West will make one trick with the ace. Meanwhile, you will take four tricks – two with the king and queen, two with low cards.

Let's see how you might use one of these plays in the context of a full deal.

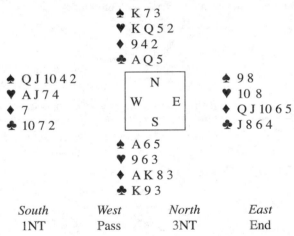

♠ K 7 3
♥ K Q 5 2
♦ 9 4 2
♣ A Q 5

♠ Q J 10 4 2   ♠ 9 8
♥ A J 7 4   ♥ 10 8
♦ 7   ♦ Q J 10 6 5
♣ 10 7 2   ♣ J 8 6 4

♠ A 6 5
♥ 9 6 3
♦ A K 8 3
♣ K 9 3

South	*West*	*North*	*East*
1NT	Pass	3NT	End

South opens 1NT. (We will discuss this bid in Chapter 5. Here it shows a flat hand with 12–14 points.) North also has a good hand and 'raises to game'. He bids 3NT.

West leads ♠ Q and you now pause for a moment to calculate how you can make the required nine tricks. You can see two certain tricks in spades, two in diamonds, and three in clubs. That is seven easy tricks with top winners. To bring the total to nine, two tricks will be needed from the heart suit.

Your plan will be to lead twice towards the high cards in dummy, hoping that West holds the missing ace.

You win the spade lead with the ace and lead a heart. West plays low and dummy's king wins the trick. It would be no good leading the second round of hearts from dummy; you must return to your hand, to lead towards ♥ Q. The ♣ 5 is led from dummy and you win the trick with the king. You now lead a second round of hearts. It makes no difference whether West plays the ace or not. You will make a second heart trick with the queen and that will bring your total to nine. 3NT bid and made.

The simple finesse

Sometimes the honour card that you are hoping will score an extra trick is accompanied by a higher non-touching honour, this type of position:

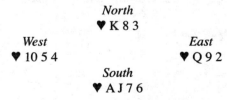

```
              North
            ♠ A Q 6
  West                    East
♠ K 10 4 3              ♠ J 9 5
              South
            ♠ 8 7 2
```

The ace is certain to score a trick, but you would like to score a second trick with the queen. As usual, you must lead *towards* the queen. You lead the 2 and, when West follows low, you play the queen. Luck is with you on this occasion. Since West held the missing king, the queen wins a trick.

This extremely common manoeuvre is known as a 'finesse'. You took a successful finesse of the queen of spades.

There are many different finessing positions.

```
              North
            ♥ K 8 3
  West                    East
♥ 10 5 4                ♥ Q 9 2
              South
            ♥ A J 7 6
```

Here you are certain to make tricks with the king and ace; you hope to score an extra trick with the jack. The king wins the first round of the suit and you next lead ♥ 3, playing towards the honour that you hope to make.

East produces the 9 on the second round and you play the jack, winning the trick. Even more good luck is to come. When you play the ace on the third round, both defenders follow. The last card in the South hand is now good. You have made four tricks from the suit.

Here you take a finesse against the jack:

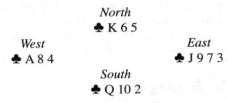

North
♣ K 6 5

West
♣ A 8 4

East
♣ J 9 7 3

South
♣ Q 10 2

You are certain to make one trick with the king or queen. You hope to make a second trick with the 10. You play low to the king, winning the first trick. Next you lead dummy's 5 and play the 10 from the South hand. The 10 forces West's ace and you make two tricks from the suit.

There is a slightly different type of finesse, which involves leading a high card to a trick. Look at this position:

North
♦ A 8 3

West
♦ K 6 4

East
♦ 9 7 5 2

South
♦ Q J 10

Do you see how to make three tricks from this diamond suit? You lead the queen on the first round. If West does not play the king, you play a low card from dummy. Because West holds the king, your queen will win the trick. You now lead the jack. If at any stage West plays the king, you will win it with dummy's ace. You score three tricks from the suit.

This position is exactly the same:

North
♦ A Q 3

West
♦ K 6 4

East
♦ 9 8 7 5

South
♦ J 10 2

You lead the jack on the first round. If West covers with the king, you win with the ace and make two more tricks with the queen and 10. If West does not cover the jack, you will lead to the queen on the next round. Three tricks either way.

Double and combination finesses

When you play for two cards to be favourably placed, you are said to take a 'double finesse'. Here you hope to catch the king and the jack:

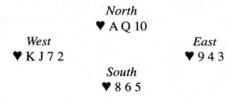

North
♥ A Q 10

West   *East*
♥ K J 7 2   ♥ 9 4 3

South
♥ 8 6 5

You begin by leading low towards the 10. Since West holds both the missing honours, the 10 wins the trick. You then return to your hand (in some other suit) and play towards the queen. Because the cards were so luckily placed, you could make three tricks from the suit. Had East held the king and jack, both finesses would have failed; you would have made only one trick. Much of the time the defenders would hold one honour each. One finesse would win, the other would fail; you would score two tricks.

Here you are missing the ace and queen:

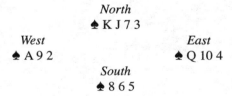

North
♠ K J 7 3

West   *East*
♠ A 9 2   ♠ Q 10 4

South
♠ 8 6 5

You lead first towards the jack. This first shot is unsuccessful, East winning with the queen. When you regain the lead, you play towards the king. Better luck this time. The ace is 'onside' (as bridge players say) and dummy's king wins the trick. When you play a third round the good luck continues. The defenders' cards break 3–3 and you score two tricks from the suit.

When two cards of equal rank are missing, you take a 'combination finesse'. This is a common position:

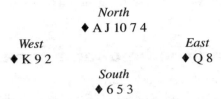

North
♦ A J 10 7 4

West
♦ K 9 2

East
♦ Q 8

South
♦ 6 5 3

A finesse of the 10 loses to the queen. On the next round a successful finesse of the jack brings in the rest of the suit. You will make four diamond tricks.

Here you are missing the ace, queen and jack:

North
♣ 8 5 3

West
♣ A J 4

East
♣ Q 7 6 2

South
♣ K 10 9

You lead towards the 10, which loses to West's jack. On the next round you play to the 9, forcing West's ace. The king will now score a trick.

Do you see why this was the best way to play the suit? You would make a trick if East held the queen, the jack, or both those cards. The alternative play of leading towards the king on the first round would yield a trick only when East holds the ace.

Points to remember

1　When you are missing one or more top cards in a suit, you can knock out the defenders' stoppers and establish tricks in the suit. A holding of ♦ K 10 6 3 opposite ♦ Q J 5 will be worth three tricks, once you have knocked out the ace.

2　You can often create extra tricks by leading *towards* high cards. For example, you lead towards a king, hoping that the ace lies to the left of the king.

3　Another example of this technique is the 'finesse'. You lead towards a combination such as A Q and play the queen in the hope that the next player does not hold the king.

Quiz

a How many tricks can you establish from ♣ Q J 9 7 opposite ♣ 10 5 3?

b If the cards lie at their most favourable, how many tricks will you make from ♠ A Q J 6 opposite ♠ 5 4 2?

c If the cards are unfavourably placed, what is the lowest number of tricks you could make from ♠ A Q J 6 opposite ♠ 5 4 2?

d How will the opponents' diamonds need to divide for you to make four tricks from ♦ A 8 7 5 3 opposite ♦ K 6 2?

Answers

a After knocking out the ace and the king, you will have two club tricks.

b If the defender in front of the ♠ A Q J 6 holds three cards including the king, you can take two successful finesses and will score four tricks.

c If the king sits over the ♠ A Q J 6 and the suit does not break 3–3, you will score only two tricks from the suit.

d They will need to divide 3–2. By losing one round, you will be able to set up four diamond tricks.

3 | THE PLAY AT NO-TRUMPS

In the previous chapter we saw some of the basic combinations within a single suit, and how to estimate the number of tricks that could be created. It is now time to 'put it together', to see how to plan the play of a whole hand.

In this chapter you will play two different hands in 3NT. On each of them you will need to ask yourself: 'How can I make nine tricks before the defenders make five?' The general idea will be to count the certain winners, then consider how to develop the extra tricks you need for the contract.

Establish the long suit

The play at no-trumps often develops into a race between the two sides to establish their respective long suits. The defender on lead normally begins with his longest suit. When you gain the lead, as declarer, you will usually start to develop your own longest suit. That's what happens on our first deal:

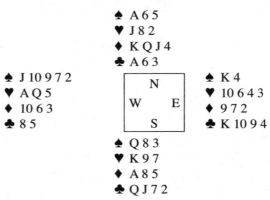

```
                    ♠ A 6 5
                    ♥ J 8 2
                    ♦ K Q J 4
                    ♣ A 6 3
    ♠ J 10 9 7 2         N          ♠ K 4
    ♥ A Q 5                          ♥ 10 6 4 3
    ♦ 10 6 3       W         E       ♦ 9 7 2
    ♣ 8 5                S           ♣ K 10 9 4
                    ♠ Q 8 3
                    ♥ K 9 7
                    ♦ A 8 5
                    ♣ Q J 7 2
```

South is to play in 3NT. If at the moment you have a problem in following the play from a diagram, take a pack of cards and arrange them as shown. You can then play the cards one at a time as we describe each trick.

Trick 1 West leads ♠ J. From a sequence of touching high cards (such as K Q J or Q J 10) it is normal to lead the highest card. You play low from dummy, because by doing so you can guarantee two spade tricks (one with the ace, another with the queen). East wins with the king.

Trick 2 Just as declarer is trying to make nine tricks, so are the defenders trying to make five. Their best chance of this here is to establish the spades, West's best suit. East therefore returns ♠ 4, won by your queen.

How many certain tricks do you have at this stage? You have two in spades, none in hearts, four in diamonds and one in clubs. That is only seven. How can you create two more, to bring the total to nine? The answer lies in the club suit. You can cash the ace ('cash' means to score a trick with a high card), then lead twice towards the queen-jack. You will make three club tricks if East has the king of clubs, or if West holds the king of clubs but the suit divides 3–3.

Trick 3 You play a club to dummy's ace,

Trick 4 ... and lead a club towards your hand. East plays low and the queen wins the trick.

Trick 5 You cross to dummy by leading a low diamond to the king ('cross' means to move from one hand to the other)

Trick 6 ... and lead a third round of clubs towards your hand. This time East rises with the king, winning the trick.

Trick 7 East leads a low heart and West wins with the queen.

Trick 8 West leads a third round of spades, knocking out dummy's ace and establishing two spade tricks for himself.

Trick 9 You cross to your hand with ♦ A.

Trick 10 You make a trick with ♣ J, discarding a heart from dummy (not a diamond because the diamonds are winning cards).

Trick 11 You play a diamond to the queen, winning the trick.

Trick 12 The jack of diamonds makes a trick too, your ninth.

Trick 13 A heart is played, won by West's ace.

As you see, it was a race between the two sides. By establishing your club suit before the defenders could enjoy their long spades, you managed to win the race on this occasion.

Perhaps the mechanics of winning tricks and crossing from one hand to another still seem very awkward to you. It's the same with anything, the first time you try it. When you are learning to drive a car it is tiresome having to manipulate the clutch, the gear lever, the brakes and the steering wheel. It soon becomes familiar and you then think little about it.

Combining two chances

Sometimes there is more than one opportunity to create the tricks you need. You must then choose which is the better prospect. On some hands you can even combine two different chances.

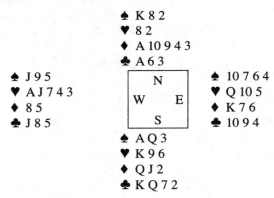

```
                    ♠ K 8 2
                    ♥ 8 2
                    ♦ A 10 9 4 3
                    ♣ A 6 3
   ♠ J 9 5          ┌─────────┐      ♠ 10 7 6 4
   ♥ A J 7 4 3      │    N    │      ♥ Q 10 5
   ♦ 8 5            │  W   E  │      ♦ K 7 6
   ♣ J 8 5          │    S    │      ♣ 10 9 4
                    └─────────┘
                    ♠ A Q 3
                    ♥ K 9 6
                    ♦ Q J 2
                    ♣ K Q 7 2
```

Again you are South, playing in a contract of 3NT.

West leads ♥ 4. When you do not have a sequence of honours, it is normal to lead your fourth-best card. The 2 is played from dummy and East plays the queen, his highest card, in an attempt to win the trick. You beat this with your king.

How many certain tricks do you have? At this stage you can count eight tricks – three spades, one heart (already made), one diamond, and three club tricks. How should you try to make a ninth trick?

One possibility is to take a finesse in diamonds, leading ♦ Q and playing low from dummy if the king fails to appear from West. This is a dangerous play, however. If West does hold the missing king, all will be well; you will score five diamond tricks, making at least 12 in all. In fact, the finesse is destined to fail. East will win with the king and return a heart, allowing his partner to make four heart tricks, thereby beating the contract.

Another possibility for the ninth trick is that the clubs break 3–3. Suppose you make three tricks with the ace, king and queen of clubs and it turns out that the defenders started with three clubs each. The fourth club in the South hand will then be good. It will give you a ninth trick.

It is possible to combine your chances in clubs and diamonds. You should test the clubs first, playing three rounds. If the suit does not break 3–3, you would then take the diamond finesse. On this particular hand the clubs do break evenly. You can therefore score the nine tricks you need without risking the dangerous diamond finesse.

Points to remember

1 The play in no-trumps is often like a race. The defenders try to establish their own longest suit. Meanwhile, declarer must try to establish enough extra tricks to make his contract.

2 When you plan the play in a no-trump contract, the first step is to count your certain tricks. You then look for the safest way to create the extra tricks you need.

4 THE PLAY IN A SUIT CONTRACT

The presence of a trump suit increases the number of ways in which tricks can be made. For example, if dummy has no cards left in one of the side suits, you can lead that suit and make a trick by ruffing. Trumps are useful also to stop the opponents from making tricks in a side suit. As soon as you have no cards left in the suit, either in your own hand or the dummy, you can ruff.

Should I draw trumps straight away?

The first big decision to make in a suit contract is whether to draw trumps immediately. You should usually do so unless you have a good reason not to. Look at this hand, played in the game contract of Four Spades.

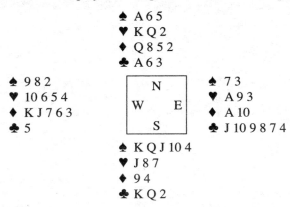

♠ A 6 5
♥ K Q 2
♦ Q 8 5 2
♣ A 6 3

♠ 9 8 2
♥ 10 6 5 4
♦ K J 7 6 3
♣ 5

N
W   E
S

♠ 7 3
♥ A 9 3
♦ A 10
♣ J 10 9 8 7 4

♠ K Q J 10 4
♥ J 8 7
♦ 9 4
♣ K Q 2

West leads the singleton ♣ 5. There are five top tricks in the trump suit and three in clubs. That is eight tricks and an extra two can be established in hearts, by knocking out the defenders' ace.

If you are unwise enough to play on hearts straight away, a disaster will occur! East will win with ♥ A and lead a club, ruffed by West. A diamond will be played to East's ace and he will give West another club ruff. West will then make a trick with ♦ K. The end effect? You will go two down in your contract of Four Spades.

You should draw trumps before playing on hearts. You win the club lead in either hand, then remove the defenders' trumps by playing three rounds of the suit. You can then play on hearts, scoring two tricks there after East has taken his ace. That will give you a total of ten tricks. The contract will be made.

If you asked a bridge player how to play Four Spades on a club lead, the reply would be: 'Win the club lead, draw trumps, and knock out the ace of hearts.'

Taking ruffs in the dummy

The main occasion when it is not advisable to draw trumps straight away is when you hope to make extra tricks by ruffing in the dummy. That's the case here:

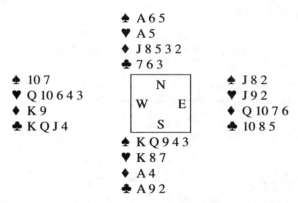

```
              ♠ A 6 5
              ♥ A 5
              ♦ J 8 5 3 2
              ♣ 7 6 3
♠ 10 7                          ♠ J 8 2
♥ Q 10 6 4 3      N             ♥ J 9 2
♦ K 9          W     E          ♦ Q 10 7 6
♣ K Q J 4         S             ♣ 10 8 5
              ♠ K Q 9 4 3
              ♥ K 8 7
              ♦ A 4
              ♣ A 9 2
```

West leads ♣ K (top of an honour sequence) and you win with the ace. How many top tricks can you count? If the trumps break 3–2, as you must hope, you will have five trump tricks. Two hearts, one diamond, and one club will bring the total to nine, one trick short of the target.

A tenth trick can be made by ruffing the third round of hearts in the dummy. This will not be possible if you make the mistake of drawing three rounds of trumps first! Dummy would then have no trumps left.

This is how the play should go:

Trick 1 You win the club lead with the ace,

Trick 2 ... then play a heart, winning with dummy's ace.

Trick 3 A second round of hearts is won with the king.

Trick 4 You lead a third round of hearts, ruffing with ♠ 5,

Tricks 5–7 ... and draw trumps in three rounds.

Trick 8 You cash ♦ A,

Tricks 9–10 ... and score two more tricks with the remaining trumps.

Tricks 11–13 The defenders make the last three tricks.

If you asked a bridge player how to play this hand in Four Hearts on a club lead, the reply would be: 'I can't afford to draw trumps straight away. I will play the two top hearts, ruff a heart, then draw trumps and cash the other winners.'

Taking discards

There is one other important occasion when you should not draw trumps straight away. It is when you need to take a quick discard before the defenders gain the lead.

The next deal is the first slam we have seen. As South, you are playing in Six Diamonds and can therefore afford to lose only one trick.

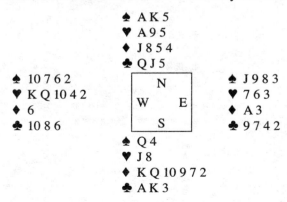

West leads ♥ K and you win with dummy's ace. What will happen if you play on trumps immediately? You will go one down. East will win with the ace and play back a heart to partner's queen.

Before allowing the defenders to gain the lead with the ace of trumps, you must dispose of your remaining heart. You can do this by playing three rounds of spades, throwing a heart on the third round. This is how the play will go:

Trick 1 West leads ♥ K, won with dummy's ace.

Trick 2 You play a low spade from dummy, winning with the queen.

Trick 3 Another spade is led, won with the ace in dummy.

Trick 4 You play dummy's king of spades, throwing your last heart.

Trick 5 Now the time has come to play trumps! You lead a trump from dummy.

Whether or not East plays the ace of trumps on this trick, that card will give the defenders their only winner.

If we asked our friendly bridge player how to play Six Diamonds after the lead of ♥ K, the reply would be: 'I can't afford to play on trumps or they'll cash a heart. I'll play three rounds of spades and throw my heart loser. Then I can play on trumps and lose just one trick to the trump ace.'

We have covered the basic ground, as far as the play of the cards is concerned. In the next few chapters we will turn to the bidding. How on earth can two players discover how many tricks they can take, just by making a few bids? They can! And we are about to find out how.

Points to remember

1 Unless there is some reason to the contrary, your first task in a suit contract is to draw trumps. This will prevent the defenders from scoring any ruffs.

2 The two main reasons why you might choose not to draw trumps immediately are: you need to take one or more ruffs in dummy, or there is an urgent need to take a discard.

5 | OPENING 1NT AND RESPONSES

The purpose of the bidding is twofold. You must find the best trump suit (or decide to play in no-trumps); you must also discover if the partnership has enough strength to play for game.

You will recall that there are 40 points in the pack. If one side has around 25 points between them, they should be able to make game. Many of the bids you can make tell your partner how many points you hold, to within a point or two. Once he knows that, he will be able to add his points to yours and decide whether it is enough for game.

The bids which define the strength of a hand quite closely are known as 'limit bids'. The most important such bid is an opening bid of 1NT. It tells your partner that you have a balanced hand (no singleton or void, usually no five-card suit). It tells him also how many points you hold, within a narrow range.

Unfortunately, we now encounter a small problem. In the UK nearly everyone plays that an opening 1NT bid shows 12, 13 or 14 points, the so-called 'weak no-trump'. Everywhere else, in particular in the USA, players prefer to play the 'strong no-trump', which shows 15, 16 or 17 points. This is a fact of life and there is no way around it. We will now describe both methods. You should read one or other section, according to where you will be playing.

The weak no-trump (UK)

Any time you hold a balanced hand of 12–14 points you should open 1NT. Your partner will have a very clear picture of your hand and will be able to tell immediately, on most hands, whether a game should be bid. These are typical hands for a weak 1NT opening bid:

(1) ♠ A 6 5   (2) ♠ K Q 7   (3) ♠ 10 6 5
     ♥ K Q 2   ♥ A Q 6 2   ♥ A J 7
     ♦ 10 9 6 2   ♦ K 10 5 4   ♦ A 4
     ♣ Q J 3   ♣ 7 6   ♣ A 9 7 6 2

Hand (1) is a minimum hand for a 1NT opening, just 12 points. The shape is fine, as you see, balanced with no singleton. Hand (2) is a maximum for 1NT, 14 points. You would also open 1NT on hand (3). You are allowed to have a five-card minor suit, if the suit division (shape) is 5-3-3-2. You would not usually open 1NT with a five-card major suit, however.

Responding to a weak 1NT

As always, the prime objective in bidding is to determine whether you have enough strength to bid game. Since you need 25 points between the two hands and your partner's 1NT opening has shown 12–14 points, you will need 11 points or more to investigate a game.

We will look first at how you should respond to 1NT when you are not strong enough to try for game.

(1) ♠ Q 7 2   (2) ♠ 10 7
     ♥ 10 8 4   ♥ K Q 6 5 2
     ♦ A 9 6 2   ♦ J 10 5 4
     ♣ K 9 3   ♣ 7 6

Hand (1) contains only nine points, so there is no chance of game. Since you have no long suit to make trumps, you would pass. No chance of game on (2) either, but you do have a five-card suit; you respond 2♥. This is a weak response, known as a 'weakness take-out', and your partner will not bid again.

When you have 11 or 12 points, there may be game if your partner has more than his minimum 12 points. You invite game by responding 2NT.

(3) ♠ K 10 3   (4) ♠ 7
     ♥ A 4   ♥ K 9 6
     ♦ Q J 9 3   ♦ A Q 9 5 4
     ♣ J 10 8 2   ♣ Q 10 7 6

On (3) you hold 11 points, also two 10s which may prove useful. You respond 2NT and your partner will then bid 3NT unless he has only 12 points or a poor 13. You would make the same 2NT response on (4), despite the weakness in spades. Do not give any thought to a game in diamonds. It is much harder to make the 11 tricks needed for a minor-suit game than nine for a no-trump game.

When you have enough strength to justify playing in game, you can either bid a game directly or jump in a suit of at least five cards.

(5) ♠ A K 3	(6) ♠ 7	(7) ♠ K J 8 6 2
♥ 9 4	♥ A Q 10 8 6 2	♥ 4
♦ A Q 9 3	♦ K 7 3	♦ A J 8 2
♣ 10 8 4 2	♣ K J 4	♣ A Q 7

On (5) you have 13 points, giving your side at least 25 points. Respond 3NT, a game contract.

On (6) you again have enough for game. When you have an eight-card fit (or better) in a major suit, you should make this suit trumps rather than play in no-trumps. With six hearts in your hand and at least two in your partner's hand, you are sure of an eight-card fit. Respond 4♥, a game contract.

On hand (7) you want to play in game but do not yet know if you have an eight-card spade fit. You respond 3♠. This is 'forcing' (in other words, your partner must bid again). Your partner will bid 3NT, holding only two-card spade support, 4♠ otherwise.

The 1NT rebid

When your hand is balanced but slightly too strong for a weak no-trump (you hold 15–17 points), you open one of a suit and 'rebid' in no-trumps.

```
♠ A 4
♥ A Q 8 3
♦ K 10 7 3
♣ K 10 5
```

This hand has 16 points, too strong for a weak 1NT. In the UK players are generally willing to open 1♥ or 1♠ when they hold only four cards in the suit, rather than the five needed elsewhere in the world. You would open 1♥. If your partner responds 1♠ you would rebid 1NT, showing a balanced hand with 15–17 points. If instead your partner responds 2♣ or 2♦, you would again rebid in no-trumps at the minimum level, now 2NT. The meaning would be the same: 15–17 points and a balanced hand. Whenever you hold a hand that is balanced, you should let your partner know by bidding no-trumps as soon as possible.

If you have already decided to play the weak no-trump, there is no need to read the next section. You can turn to 'The Stayman convention' on page 37.

The strong no-trump (USA)

Any time you hold a balanced hand of 15–17 points you should open 1NT. Your partner will have a very clear picture of your hand and will be able to tell immediately, on most hands, whether a game should be bid. These are typical hands for a strong 1NT opening bid:

(1)	♠ A 6 5	(2)	♠ K Q 7	(3)	♠ A 9 7 6 4
	♥ K Q 2		♥ A Q 6 2		♥ A J 7
	♦ K 9 6 2		♦ A Q 10 5		♦ A 4
	♣ Q J 3		♣ 7 6		♣ Q 9 5

Hand (1) is a minimum hand for a strong 1NT, just 15 points. The shape is fine, as you see, balanced with no singleton. Hand (2) is a maximum for the 1NT opening, 17 points. Do not be deterred from the bid because you have no 'stopper' (high card to prevent your opponents from running a suit) in clubs.

You would also open 1NT on hand (3). When you are in the 15–17 point range, it is generally best to open 1NT on hands with a 5-3-3-2 shape. This applies even when the five-card suit is a major.

Responding to a strong 1NT

As always, the prime objective in bidding is to determine whether you have enough strength to bid to game. Since you need 25 points between the two hands and your partner's 1NT opening has shown 15–17 points, you will need eight points or more to investigate a game.

We will look first at how you should respond to 1NT when you are not strong enough to try for game.

(1)	♠ Q 7 2	(2)	♠ 10 7
	♥ 10 8 4		♥ K 10 9 6 2
	♦ J 9 6 2		♦ J 10 5 4
	♣ K 9 3		♣ 7 6

Hand (1) contains only six points, so there is no chance of game. Since you have no long suit to make trumps, you would pass. No chance of game on (2) either, but you do have a five-card suit; you respond 2♥. This is a weak response, known as a 'weakness take-out', and your partner will not bid again.

When you have eight or nine points, there may be game if your partner has more than a minimum 15 points. You invite game by responding 2NT.

(3) ♠ 10 9 3   (4) ♠ 7
     ♥ A 4   ♥ K 9 6
     ♦ Q J 9 3   ♦ A 10 9 5 4
     ♣ J 10 8 2   ♣ J 10 7 6

On (3) you hold eight points, also two 10s which may prove useful. You respond 2NT and your partner will then bid 3NT unless he has only 15 points or a poor 16. You would make the same 2NT response on (4), despite the weakness in spades. Do not give any thought to a game in diamonds. It is much harder to make the eleven tricks needed for a minor-suit game than nine for a no-trump game.

When you have enough strength to justify playing in game, you can either bid a game directly or jump in a suit of at least five cards.

(5) ♠ A 7 3   (6) ♠ 7   (7) ♠ K J 8 6 2
     ♥ 9 4   ♥ A Q 10 8 6 2   ♥ 4
     ♦ A Q 9 3   ♦ 7 3 2   ♦ Q 10 8 2
     ♣ 10 8 4 2   ♣ K J 4   ♣ A Q 7

On (5) you have ten points, giving your side at least 25 points. Respond 3NT, a game contract.

On (6) you again have enough for game. When you have an eight-card fit (or better) in a major suit, you should make this suit trumps rather than play in no-trumps. With six hearts in your hand and at least two in your partner's hand, you are sure of an eight-card fit. Respond 4♥, a game contract.

On hand (7) you want to play in game but do not yet know if you have an eight-card spade fit. You respond 3♠. This is 'forcing' (in other words, your partner must bid again). Your partner will bid 3NT, holding only two-card spade support, 4♠ otherwise.

The 1NT rebid

When your hand is balanced but too weak for a strong no-trump (you hold 12–14 points), you have to make a special opening bid. You bid 1♣, pretending for a moment that you hold clubs, with the intention of rebidding 1NT.

(8) ♠ A 8 4
     ♥ Q 10 8 3
     ♦ K 9 7
     ♣ K 10 5

Hand (8) has 12 points, too weak for a strong 1NT. In the USA and most of Europe except the UK, players are not willing to open 1♥ or 1♠ on a suit of only four cards. You must therefore open 1♣ on balanced hands in the 12–14 range. If your partner responds 1♦ or 1♠, you will rebid 1NT and he will place you with exactly this type of hand. If instead he responds 1♥, you will raise to 2♥ because a fit of at least eight cards has been found.

The Stayman convention

A 'convention' is an agreement with your partner that a bid will have a special meaning – a meaning other than a suggestion that the bid suit should become trumps. The first convention we will look at is one of the most famous in the world, the Stayman convention.

We have already seen that when you have an eight-card fit in a major suit it is better to make that suit trumps than to play in no-trumps. The Stayman convention is aimed at finding a 4–4 fit in a major suit after partner has opened 1NT (weak or strong, it makes no difference). You respond 2♣. This says nothing about your clubs. It is a conventional bid that asks your partner if he holds a four-card major. He will reply along these lines:

> 2♦ No, I have no four-card major.
> 2♥ Yes, I have four hearts (and maybe four spades).
> 2♠ Yes, I have four spades.

To use the convention you must have enough strength to invite game, at least.

> ♠ 6 4
> ♥ A Q 9 3
> ♦ K Q 7 3
> ♣ Q 8 5

When your partner opens 1NT you want to be in game, but which game? If your partner holds four hearts you would prefer the heart game to 3NT. You respond 2♣, Stayman. If partner rebids 2♥ you will raise to 4♥. Otherwise you will bid 3NT.

Suppose you had the same shape of hand but only enough points to invite game. You would then raise your partner's 2♥ rebid to 3♥, meaning 'go on to game unless you have a minimum 1NT opening'. If instead your partner rebid 2♦ or 2♠, denying a four-card heart suit, you would invite a no-trump game by bidding 2NT at your next turn.

Since a response of 2♣ has this special meaning, you have to pass 1NT when you hold a weak hand with five or more clubs.

When you are balanced with 18–19 points

Whenever your hand is balanced, your aim will be to let your partner
know, by making a bid in no-trumps as soon as possible. With 18 or 19
points you open one of a suit, then *jump* in no-trumps.

♠ A J 9 2
♥ K 4
♦ A Q 7 3
♣ K Q 5

If you and your partner are willing to open 1♥ or 1♠ with only a four-card
suit (normal in the UK), you would open 1♠ here. Over a response at the
two level you would jump to 3NT. If instead you play five-card majors
(meaning that you do not open 1♥ or 1♠ unless you hold at least five cards
in the suit), then you would open 1♦. Again your plan would be to jump in
no-trumps (1♦ – 1♥ – 2NT, or 1♦ – 2♣ – 3NT).

Perhaps you are wondering what to do when you hold a balanced hand
with 20 points, or 23 or 27? There are special opening bids for such hands,
as we will see in Chapter 12.

Points to remember

1 When you hold a balanced hand, let your partner know as
   soon as possible. If the hand does not fall into your agreed
   range for a 1NT opening, plan to rebid in no-trumps.

2 If you have agreed to play the weak 1NT (12–14 points), you
   will open One of a suit on 15–17 points, planning to rebid
   no-trumps at the minimum level.

3 If you have agreed to play the strong 1NT (15–17 points),
   you will open 1♣ on balanced hands in the 12–14 point
   range, planning to rebid 1NT.

4 With 18–19 points you open One of a suit and jump-rebid in
   no-trumps (whatever opening 1NT you play).

5 In response to a 1NT opening, 2♣ is Stayman (asking for a
   four-card major); Two of any other suit is weak; Three of a
   suit is forcing and shows a suit of at least five cards.

Quiz

a What is the difference between a 1NT opening in the UK and the same opening in the USA?

b When your partner opens 1NT, what does it mean if you respond 2♣?

c When your partner opens 1NT, what does it mean if you respond 2♠?

d When your partner opens 1NT, what does it mean if you respond 3♥?

e If the bidding goes 1♣ – 1♥ – 1NT, how many points does the 1NT rebid show?

f If the bidding goes 1♣ – 1♥ – 2NT, how many points does the 2NT rebid show?

Answers

a In the UK most players play a weak 1NT of 12–14 points. In the USA (and in the rest of Europe) most players prefer to play a strong 1NT of 15–17 points.

b A response of 2♣ is the Stayman convention, asking whether partner holds a four-card major suit.

c A response of 2♠ is weak (whatever range of 1NT you play). The opener should then pass.

d A response of 3♥ is strong and shows at least five hearts. The opener will usually rebid 3NT with only two hearts, otherwise 4♥.

e If you play the weak no-trump a 1NT rebid shows 15–17 points. If you play the strong no-trump a 1NT rebid shows 12–14 points.

f A jump rebid of 2NT shows 18–19 points.

6 | OPENING SUIT BIDS OF ONE AND RESPONSES

In the previous chapter we saw how to open the bidding when you hold a balanced hand. Now we look at unbalanced hands, those containing a five-card suit. On most hands within the 12 to 20 point range you will open with a one-bid in your longest suit. Suppose you hold any of these hands:

(1)		(2)		(3)	
♠	6 2	♠	A J 7	♠	9
♥	A Q 9 6 2	♥	J	♥	Q 7
♦	K 10 8 3	♦	A K 10 5 4	♦	A 9 7 5 3
♣	K 3	♣	A Q J 4	♣	A K 7 6 2

Hand (1) represents a minimum opening of 1♥. Hand (2) contains 20 points; you would open 1♦, but the hand is a maximum for an opening bid at the one level. We will see in Chapter 12 what opening bid to make when your hand is even stronger. On hand (3) you have two five-card suits. In this case you should open one of the *higher* suit. Here you open 1♦ rather than 1♣, even though the club suit is stronger.

Although 12 points is the normal minimum for a one-bid, you can open with slightly fewer points when you have a long and strong suit, or two good suits.

(1)		(2)	
♠	K Q J 10 8 6	♠	9 4
♥	A 9 7	♥	A Q 10 8 4
♦	4	♦	A J 7 6 2
♣	10 8 3	♣	2

Hand (1) has a fine six-card major suit; you would open 1♠ despite holding only 10 points. You intend to rebid 2♠, showing a minimum opening with a long suit. Hand (2) has two good five-card suits. You would happily open 1♥, intending to rebid 2♦.

Against that, you would not make an opening bid on either of these hands:

(3)   ♠ Q 9 7 6 4 2   (4)   ♠ K 4
        ♥ A 3   ♥ Q 10 9 5 2
        ♦ J 3   ♦ K J 7 6 2
        ♣ Q J 5   ♣ Q

Hand (3) does contain 10 points and a six-card major, but most of the points lie in the short suits. Hands normally generate more tricks when the high cards lie in the long suits. You would pass on this hand.

Hand (4) contains 11 points and two suits. But a bridge player would describe this as a 'poor 11-point hand'. There are no aces, always a bad sign. Also, the singleton queen of clubs will not help much; it is not worth the two points normally assigned to a queen. Again you would pass.

Responding to a one-bid

Your partner opens with a bid such as 1♥ and the next player passes. How much do you need to respond? Normally six points is enough to make a bid of some sort. There are many different responses available and we will divide them into these categories:

- minimum responses (around 6–9 points)
- responses that invite a game (about 10–12 points)
- game bids (usually 13 or more points)
- wide-range responses
- responses inviting a slam (usually 16+ points).

Responding on weak hands

Suppose your partner has opened 1♥ and you hold any of these hands:

(1)   ♠ K 6 2   (2)   ♠ A J 4   (3)   ♠ J 8 6 3
       ♥ 7 2   ♥ 8 3   ♥ A 10 7 6
       ♦ J 10 8 3 2   ♦ K J 5 4   ♦ J 3
       ♣ 9 7 2   ♣ 10 8 4 3   ♣ 9 5 4

Hand (1) does not contain the necessary six points, so you would pass. Hand (2) is worth a response but (as we will see in a moment) it is not strong enough for a two-level response in clubs or diamonds. You would respond 1NT. This is a 'limit bid', telling your partner that you hold 6–9 points.

Hand (3) contains four-card support for your partner's suit and you would raise to 2♥. Again this is a limit bid. Your partner will place you with around 6–9 points. It is always a good idea to make a limit bid if you can. Your partner can then add the strength you have shown to his own; by doing so, he will know if a game is possible.

Responding with a game invitation

When your hand is in the 10–12 point range you are entitled to invite a game. You will choose a bid, or sequence of bids, that has the meaning 'I have a hand in the middle range, about 10–12 points. If we have a good trump fit or you hold more than a minimum opening, we should be able to make game.'

Suppose again that your partner has opened 1♥. There are two responses that will immediately invite a game: 2NT and a double raise of the opener's suit. Suppose you hold one of these hands:

(1)	♠ A 10 4	(2)	♠ 8 3
	♥ J 3		♥ A Q 5 2
	♦ A 9 8 4		♦ J 7 6
	♣ K 7 6 2		♣ A 10 8 4

On (1) you hold 12 points, with no support for hearts but a 'stopper' (a high card) in each of the other suits. You respond 2NT, a limit bid that shows exactly this type of hand. It is now up to your partner to decide whether a game should be bid.

Hand (2) contains 11 points, also four-card support for your partner's hearts. You would invite game by jumping from 1♥ to 3♥. Again this is a limit bid. Your partner must now decide whether to pass or advance to game.

Bidding game immediately

When your partner opens the bidding and you have the equivalent of an opening bid yourself, a game should be bid. Sometimes you can bid game directly. Suppose your partner has opened 1♠ and you hold any of these hands:

(1)	♠ A K 6 2	(2)	♠ 7	(3)	♠ 6 3
	♥ 7 2		♥ A Q J 10 7 6 2		♥ A Q 7
	♦ K J 3 2		♦ K Q 5		♦ K J 8 3
	♣ Q 8 4		♣ 10 3		♣ K 9 6 4

Hand (1) contains four-card spade support and 13 points. You want to be in game and you know there is a good trump fit in spades. Bid 4♠. On (2) you have a good suit of your own and enough strength to play in game, facing an opening bid. Respond 4♥. With hand (3) it is very likely that 3NT will be the best game. Let your partner know the good news. Respond 3NT.

Remember that the most important objective of the bidding is to bid game when you and your partner have the values to do so. When your partner has opened the bidding and you hold 13 points or more, it is your responsibility to make sure that game is bid. You must not make any 'limit bid' below game which your partner is allowed to pass. Don't be one of those players who says, 'Oh dear, we missed game. I was hoping you would bid again.'

Wide-range responses

Although it is a good idea to make an immediate limit-bid response whenever you can, on most hands this is not possible. The usual reason is that the responder does not yet know which suit, if any, should become trumps. In such a situation the responder merely bids a new suit and waits to see what the opener does next. Very often the opener will make a limit bid at his second turn and the picture will become clearer.

To respond in a new suit at the one level, you need around six points. A response at the two level will carry the bidding higher; you need about ten points to make such a move.

We will look first at some cases where the responder is relatively weak. Suppose your partner has opened 1♥ and you hold one of these hands:

(1)	♠ A J 6 2	(2) ♠ Q 4
	♥ 7 2	♥ 6 2
	♦ 9 8 4	♦ 10 8 6 2
	♣ J 10 5 4	♣ A J 9 7 3

On hand (1) you respond 1♠. You have the necessary six points to respond at the one level and you hope to find a spade fit. On hand (2) you have enough to respond, but not the ten points necessary to respond at the two level. On all such hands you must respond 1NT. This response does not necessarily show a balanced hand. It shows that you have a weakish responding hand in the 6–9 point range, and no suit that you could bid at the one level.

Let's suppose now that you have more points, enough to bid a game or at least to invite one (your partner has again opened 1♥):

(3)   ♠ A J 6 3   (4)   ♠ A J 6 3
        ♥ Q 8 3   ♥ J 7
        ♦ A 9 3   ♦ Q 8 4
        ♣ 8 7 3   ♣ A K 9 4

On hand (3) you are in the 10–12 point range and will therefore want to invite game. But which game? Your partner might have a spade fit with you. Or he might hold five or more hearts, which would make hearts a good trump suit. You should mark time with a 1♠ response. After your partner's next bid, you will have a better idea which suit if any should be made trumps; you will then be able to invite game.

On hand (4) you have enough for game, but it could be 3NT, 4♥, 4♠ or even 5♣. If your partner has a very strong hand, it might even be possible to bid a slam. You respond 1♠ for the moment. This is a wide-range response and does not limit your hand. You might hold only six points; you might sometimes hold as many as 20 points.

Perhaps you are wondering how a simple response of 1♠ on the last hand fits in with our advice that you must ensure game is bid when the values are there. The answer is that a response in a new suit is 'forcing'. Your partner must bid again, however minimum his opening bid was. You will then have the chance to bid a game, or perhaps to make another 'forcing bid'.

Which suit should you respond in when you have more than one suit? With two four-card suits, you bid the 'cheaper' suit. What do we mean by that, you may wonder. Look at this hand:

(5)   ♠ A 10 7 2
        ♥ 7 4
        ♦ A J 5 2
        ♣ 10 8 3

When your partner opens 1♣ you respond 1♦; if there is a 4–4 spade fit, your partner will now have the chance to bid the suit. If instead your partner opened 1♥, you would respond 1♠. He could then rebid 2♦, locating any fit there. That is what 'cheaper suit' means – you keep the bidding as low as possible, leaving room for your partner to bid your other suit.

When you hold two five-card suits, the rule is the same as for opening bids: you should bid the *higher* suit first.

(6) ♠ A 10 6 4 3
     ♥ K Q 7 6 2
     ♦ J 3
     ♣ 5

If your partner opens with one of a minor, you respond 1♠ with the intention of bidding 2♥ on the next round of bidding. By bidding your suits that way round you make it easy for your partner to bid 2♠ over 2♥ if he prefers your first suit.

When you hold one five-card suit and one four-card suit you respond in the longer suit, provided you have the strength to do so. Suppose your partner has opened 1♦ and you hold either of these hands:

(7) ♠ A J 6 3   (8) ♠ A 9 8 2
     ♥ 3   ♥ J 7
     ♦ J 9 3   ♦ 9 4
     ♣ A Q 9 8 2   ♣ A 10 8 6 3

Hand (7) is in the game-try region of around 11 points. You are strong enough to respond in your five-card suit, bidding 2♣. If your partner rebids 2♦ you will bid again, saying 2♠. Bidding two suits in this way, as responder, shows at least game-try strength. Hand (8) is not strong enough for such a treatment. You do have a suit that you can show at the one level, however, so you would respond 1♠ rather than 1NT.

Responses that invite a slam

We will look at slam bidding in Chapter 14. For the moment we will just say that when the responder has 16 points or more he may wish to inform his partner immediately, so that a slam can be considered. He does this by making a jump bid in a new suit, a 'jump shift' as it is called.

Suppose your partner has opened 1♥ and you hold one of these beauties:

(1) ♠ A K Q 10 6 3   (2) ♠ 9 3
     ♥ K 2   ♥ A Q 8 7
     ♦ A 3   ♦ A 3
     ♣ 8 7 3   ♣ A K J 8 7

With 16 points and a fine six-card spade suit, you have more than enough on hand (1) for game to be certain; a slam may well be possible. You announce this with a jump response of 2♠. On hand (2) you have a splendid fit for hearts and a full 18 points. You make a jump shift to 3♣, intending to show your heart fit on the next round.

One-level opening bids – points to remember

1 With an unbalanced hand you should open One of your longest suit. This opening covers a wide range, about 12–20 points.

2 With two five-card suits (or two six-card suits), open the higher suit, intending to bid the other suit on the next round.

3 When you hold a strong six-card suit, or two good five-card suits, you may open on 11 points, sometimes on only 10 points.

Responding to one-level bids – points to remember

1 There are two weak responses: 1NT and a single raise of your partner's suit. These both suggest 6–9 points.

2 There are two game-invitational responses: 2NT and a double raise of your partner's suit. These suggest around 10–12 points.

3 On many hands the best response is to introduce a new suit. This is forcing and does not limit your hand. After your partner has bid again you will have a better idea whether you should make any particular suit trumps, also whether you should try for game.

4 You need six points to respond in a new suit at the one level (1♦ – 1♠), ten points to respond in a new suit at the two level (1♦ – 2♣).

Quiz

a What opening bid would you make, if any, on these hands:

(1)	♠ A 6	(2)	♠ A K Q 4	(3)	♠ J 9 7 6 3
	♥ Q 9 7 6 2		♥ J 2		♥ A Q
	♦ A Q 9 5 4		♦ 9 5		♦ J 8 3
	♣ J		♣ Q 7 6 3 2		♣ K 6 4

b When partner has opened 1♠, how many points would a response of 1NT show?

c Suppose your partner has opened 1♥. What would you respond on each of these hands:

(1)   ♠ A 10 8 3   (2)   ♠ A J 6 4   (3)   ♠ J 9 7
      ♥ 9 4   ♥ 7 2   ♥ A Q 9 4
      ♦ 5 2   ♦ Q 5   ♦ A 10 3
      ♣ Q J 8 7 3   ♣ A Q 9 3 2   ♣ 8 7 5

d When your partner has opened 1♥, which of these responses force him to bid again: 1♠, 1NT, 2NT, 3♥?

e Your partner opens 1♣. What response would you make on these hands:

(1)   ♠ J 7 5 2   (2)   ♠ A 5   (3)   ♠ K 10 3
      ♥ Q 6 4   ♥ Q 10 5 3 2   ♥ Q 7 2
      ♦ A Q 10 3   ♦ A K 7 4 2   ♦ 9 2
      ♣ 7 3   ♣ 6   ♣ A Q 9 8 3

Answers

a (1) Open 1♥. With two five-card suits open the higher suit.

   (2) Open 1♣. With two suits of different length, open the longer suit.

   (3) Pass. Only 11 points and the spades are poor.

b   6–9 points.

c (1) Respond 1♠. You are not strong enough to respond at the two level.

   (2) Respond 2♣. You are strong enough to show the clubs, then bid the spades on the next round.

   (3) Raise to 3♥, inviting a heart game.

d   Only 1♠ is forcing. The other three responses are limit bids.

e (1) Respond 1♦, making the cheaper bid with four-card suits.

   (2) Respond 1♥, bidding the higher of two five-card suits.

   (3) Respond 3♣, showing about 10–12 points.

7 | THE OPENER'S REBID

We look now at the opener's second bid. When the responder made a limit bid, the opener may already be in a position to name the final contract. When instead the responder made a wide-range response, such as bidding a new suit, it will be the opener's responsibility to show his own strength. Wherever possible he will do this by making a limit bid himself, leaving the responder to set the final contract.

There are many areas to cover and we will deal with them in this order:

■ partner has responded 1NT
■ partner has responded 2NT
■ partner has given a single raise
■ partner has given a double raise
■ partner has bid a new suit at the one level
■ partner has bid a new suit at the two level
■ partner has made a jump shift.

Partner has responded 1NT

Any time your partner has made a limit bid, you add your own strength to that shown by him. A 1NT response shows 6–9 points. Remembering that around 25 points will be needed for game, what would you rebid on these hands after a start of 1♥ – 1NT?

(1)   ♠ 8   (2)   ♠ A Q 7   (3)   ♠ K Q 7 6
           ♥ A Q 9 6 2   ♥ K Q 8 7 3   ♥ A J 9 8 3
           ♦ A 8 4   ♦ 10 2   ♦ A J 4
           ♣ K 10 7 5   ♣ A Q 4   ♣ 5

Even if your partner has a maximum nine points, there will not be game available on (1). You should rebid 2♣, showing your second suit. Your partner can then pass, if he likes clubs as a trump suit, or bid 2♥. Either way, you are likely to end in a better contract than 1NT, where the spades may be poorly protected.

Hand (2) contains 17 points and game may be possible if your partner has a better than minimum 1NT response. You therefore invite game by raising to 2NT. You would do the same with 18 points. Holding 19 points, you would jump to 3NT, confident that 25 points were present.

You must be careful on hand (3). There are two good reasons why you should not rebid 2♠. Firstly, if your partner prefers to play in hearts he will have to go to 3♥, dangerously high. Secondly, your partner has already denied a four-card spade suit by responding 1NT. There is no point in looking for a spade fit and you should pass 1NT.

Partner has responded 2NT

This response, you will recall, shows around 11–12 points. The opener will pass only with a minimum opening bid and no particularly long suit. Suppose the bidding has started 1♥ – 2NT and you hold one of these hands:

(1)	♠ 8 5 3	(2)	♠ A 7	(3)	♠ 7 6 3
	♥ A Q 9 6 2		♥ K Q 10 7 3		♥ A J 9 8 6 3
	♦ Q 10 3		♦ A J 8 3		♦ 4
	♣ A 7		♣ 9 3		♣ K Q 7

On (1), a bare 12 points, you would pass 2NT. If ♦ Q were ♦ K, you would raise to 3NT. On (2) you would rebid 3♦, forcing for one round. If your partner rebids 3♥, you will raise to the heart game; otherwise you will play in 3NT. Hand (3) has only 10 points. You would rebid 3♥, a weak call; you expect your partner to pass unless he has undisclosed heart support.

Partner has made a single raise

When your partner has raised your major-suit opening to the two level, the hand opposite will be in the 6–9 point range. After a start of 1♠ – 2♠, what would you say on:

(1)	♠ A Q J 8 5	(2)	♠ A K 10 4 2	(3)	♠ K Q 7 6 3
	♥ 10 5		♥ 4		♥ A Q 9 8 3
	♦ K 3		♦ K 7 6		♦ A 4
	♣ K 9 4 2		♣ A J 8 2		♣ 5

Hand (1) is not strong enough to try for game. Unless you have a very distributional hand (long suits, singletons or voids), you need around 16 points to try for game.

Hand (2) is worth a game try. The best bid is 3♣. This is not a suggestion to make clubs trumps (you have already decided to play in spades); it is a 'trial bid', telling your partner that game is possible. You make a trial bid in your second longest suit, to give your partner a good picture of your hand. If he is short in clubs, for example, he will know that you can score some ruffs in the short trump hand; he will be more inclined to bid game.

Hand (3), with 15 points and two good five-card suits, is worth a rebid of 4♠. When you have enough to bid game, there is no point in bidding your other suit.

You will gather from the above examples that when you are considering bidding game in a suit, high-card points are not everything. Distribution (the shape of the hand) is important also. If you like to measure everything in terms of points, a second five-card suit is worth an additional three points. So, hand (3) above counts as 18 points, enough for game facing 6–9 with partner.

What if your partner has made a single raise in a minor suit? Game in clubs or diamonds requires 11 tricks, always a distant target. It is more likely that you can make 3NT. Suppose the bidding has started 1♦ – 2♦ and you have to find a rebid on one of these hands:

	(4)	♠ A Q 8 5	(5)	♠ A 10 4
		♥ 10 4		♥ K J 3
		♦ A K J 8 7		♦ K Q 7 5 2
		♣ A 5		♣ A 8

Hand (4) contains 18 points, which puts you in the game zone opposite your partner's 6–9. You will not be able to make 3NT, though, unless your partner has something good in hearts. You should rebid 2♠. This is not so much a suggestion to make spades trumps (your partner would probably have responded 1♠ with four spades), it tells your partner that you are strong enough to consider a game and have the spades well protected.

On hand (5) you have 17 points and all the suits stopped. You suggest a game by rebidding 2NT. If your partner is in the top half of the 6–9 points range, he will raise to 3NT.

Partner has made a double raise

Similar considerations apply when your partner has given a double raise. Of course, you will now need less yourself to advance to game. How would you judge these hands after a start of 1♥– 3♥?

(1)   ♠ Q 3
       ♥ A Q 8 7 2
       ♦ K 8 3
       ♣ Q 9 4

(2)   ♠ 9 2
       ♥ A K 10 7 6 2
       ♦ K J 6 3
       ♣ 2

On (1) you would pass your partner's double raise. You have a fairly flat hand with only 13 points; also, the queen doubleton in spades is not worth much.

Hand (2) contains only 11 points, but just look at the shape! A six-card suit and a singleton. You would raise to 4♥.

In case you are in any doubt about shape being as valuable as points, let's put each of these hands opposite a typical double raise. This is the first one:

West   *East*
♠ Q 3   ♠ A 6 4
♥ A Q 8 7 2   ♥ J 9 5 3
♦ K 8 3   ♦ Q 7 2
♣ Q 9 4   ♣ K 10 8

Plenty of tricks to lose, aren't there? You are likely to lose a spade trick and possibly a trump trick. In the minor suits, you are certain to lose to the two aces and might in fact lose two tricks in each minor. That West hand is not really worth 13 points, as you see.

On the second hand, prospects are brighter:

West   *East*
♠ 9 2   ♠ A 6 4
♥ A K 10 7 6 2   ♥ J 9 5 3
♦ K J 6 3   ♦ Q 7 2
♣ 2   ♣ K 10 8

Game looks very good. You expect to lose one spade, no trumps (unless you are unlucky and North holds all three outstanding trumps), the ace of diamonds and the ace of clubs. So, two points fewer than the first hand but much of the time you would score two tricks more.

Some beginners' books give a set of rules: add two points for a singleton, one point for this, three points for that... That's not how good players assess a hand. They would describe hand (1) as: a poor 13-count, flat shape, and two unsupported queens. (The term 'unsupported' means that there is no other honour in the suit; honours tend to be more productive when accompanied by other honours.) Hand (2) would be described as: an excellent 11-count, good trumps, and diamond side suit containing two honours.

Partner has responded in a new suit at the one level

Suppose the auction starts 1♦ – 1♠. How much is known about the strength of the two hands? Very little. Both players have made wide-range bids rather than limit bids. The opener may have 12 points or may have 20 points. Similarly, the responder may have a bare six points, or he may have 20 points. The players have named one possible trump suit each and that is all for the moment.

After such a start the opener will often be the first to make a limit bid. There are three ways of doing this: you can rebid in your own suit, you can raise your partner's suit, or you can rebid in no-trumps. In each case you will give a picture of the strength of your hand. Basically, the higher you bid, the stronger your hand will be.

We will look first at the cases where, as opener, you rebid your own suit. Suppose the auction has started 1♦ – 1♠ and you must find a rebid on one of these hands:

	(1)	♠ J 3	(2)	♠ 5
		♥ A 8 5		♥ Q 10 3
		♦ A Q 10 7 6 2		♦ A K J 6 4 3
		♣ J 5		♣ A Q 3

The first hand is the minimum range for an opening bid. You would rebid 2♦, showing at least five diamonds and around 12–14 points. The second hand is stronger. You would rebid 3♦, showing at least six diamonds and a hand in the middle range (15–17 points).

The same idea – the higher you bid, the stronger you are – is followed when you raise your partner's suit. Suppose again that the bidding has started 1♦ – 1♠ and you must seek a good rebid on one of these hands:

(3)	♠ Q 10 8 5	(4)	♠ A 9 8 3	(5)	♠ K Q 5 2
	♥ A 5		♥ 4		♥ A 4
	♦ A Q 8 4 2		♦ K J 7 6 2		♦ A K Q 7 6
	♣ 10 7		♣ A K 5		♣ 8 5

Hand (3) is in the minimum range, with no particularly good shape. You would raise to 2♠. Hand (4) is in the middle range, with better shape too. You would raise to 3♠. Hand (5) is in the top range of opening bids (18–20). Even though the shape is unexciting, you have enough to raise to 4♠, a game contract.

We have already discussed the situation of rebidding in no-trumps (see Chapter 5). You will recall that in a sequence such as 1♦ – 1♠ – 1NT, the 1NT rebid shows a hand of 15–17 points if you play the weak no-trump, a hand of 12–14 points if you play the strong no-trump. A jump rebid in no-trumps shows 18–19 points.

The only remaining option is for the opener to bid a new suit. Suppose the bidding has started 1♥ – 1♠ and you hold one of these hands:

(6)	♠ 4	(7)	♠ 6 4	(8)	♠ K 5
	♥ A K J 7 2		♥ A K Q 8 3		♥ A K J 8 2
	♦ J 4 2		♦ K 4		♦ 4
	♣ K 10 6 3		♣ A J 5 4		♣ A K J 5 2

No problem on (6). You rebid 2♣, showing that you have a second suit of at least four cards. A rebid in a new suit is wide-range. You would rebid 2♣ also on hand (7), which is in the middle range. Only when you come to a splendid hand such as (8), in the top range, would you make a jump rebid in a new suit, here 3♣. Such a jump is 'game forcing', which means that neither partner can stop bidding until a game has been reached.

Making a 'reverse' bid

You will have gathered by now that the higher you take the bidding, the better the hand you must have. Compare these two sequences:

(1)	West	East	(2)	West	East
	1♦	1♠		1♦	1♠
	2♣			2♥	

In the first sequence East will be able to give preference to West's first suit by bidding 2♦. In the second sequence West has rebid at the two level in a suit ranking higher than his first suit. This forces the bidding higher. If East, with some miserable six-count, prefers diamonds to hearts, he will now have to go to the three level, bidding 3♦. It follows that West cannot afford to rebid beyond two of his first suit unless he is quite strong. A rebid such as 2♥ in (2) is known as a 'reverse' and requires around 17 points, or compensating shape. Suppose, after a start of 1♦ – 1♠, that you hold one of these hands:

(3)	♠ 9	(4)	♠ A 3
	♥ A Q 5 4		♥ K J 10 4
	♦ K J 7 6 2		♦ A K 10 7 4
	♣ K 5 3		♣ Q 3

Hand (3) is in the minimum range, well short of the values needed for a 'reverse'. You must rebid 2♦. Hand (4) does have the required strength. You would rebid 2♥, letting your partner know that you have a good hand. A reverse bid is forcing.

Note that your first suit will always be longer than your second suit when you reverse. Suppose you have a hand with five hearts and five diamonds. You would always open 1♥ rather than 1♦, regardless of the strength of the hand. You plan to rebid a wide-range 2♦, or a game-forcing 3♦. It is not correct to open 1♦ on such a 5–5 hand, planning to reverse.

Partner has responded in a new suit at the two level

Similar considerations apply when your partner has bid a new suit at the two level. He has shown ten points or more by such an action, however, so if you have a hand in the middle range (15–17 points), you will want to reach a game somewhere.

In each of these sequences the opener shows a hand of at least 15 points:

(1)	West	East	(2)	West	East
	1♥	2♣		1♥	2♣
	2NT			3♥	

In the first, the opener has rebid 2NT, showing 15–17 points. In the second sequence, the opener has made a jump rebid in his suit, again showing a hand of good strength. In both cases the combined values will be 25 points or more and neither partner should let the bidding stop short of game.

(Some players in the USA and Europe treat the 2NT rebid as weak after a two-level response. You would have to discuss the matter with your partner. If you do decide to play the 2NT rebid as weak, you will not need to use the special 1♣ opening on balanced hands in the 12–14 point range.)

	(3)	*West*	*East*	(4)	*West*	*East*
		1♥	2♦		1♥	2♦
		2♠			3♣	

West has reversed in (3), and made what is known as 'high reverse' (a non-jump rebid in a new suit at the three level) in (4). In both cases the bidding must continue to game.

Partner has made a jump shift

A jump shift by your partner (such as 1♦ – 2♠) suggests that a slam may be possible. At the very least the bidding must continue to game. There are no particular rules governing the rebid by the opener, except that you should not make a jump rebid. Your partner's response has already consumed an extra level of bidding, so you should show your shape, leaving any expression of additional strength until later.

Suppose after the start 1♦ – 2♠ you hold one of these hands:

(1)	♠ 4	(2)	♠ Q 7 3	(3)	♠ Q 5
	♥ A 7 2		♥ 5		♥ A Q 5
	♦ A K J 10 7 3		♦ A Q 10 4 2		♦ A K 7 5
	♣ K J 4		♣ A 10 6 3		♣ K 8 3 2

On hand (1) you would have jumped to 3♦ after a start of 1♦ – 1♠. Do not now jump to 4♦. Keep the bidding low, rebidding 3♦, and see what your partner has to say next. There will be plenty of time to move towards a slam later. Don't bother showing your moderate clubs on hand (2). Your partner will be much more interested to hear of your very respectable spade support. Rebid 3♠. The jump shift guarantees a good spade suit.

On hand (3) you would have jumped in no-trumps, had your partner made a low-level response. Now simply rebid 2NT, keeping the bidding low and waiting to hear whether your partner wants to support your diamonds or rebid his spades. With such a strong hand you intend to carry the bidding to the slam level anyway.

Points to remember

1 Where possible, choose a rebid which is a limit bid. This will
   allow your partner to judge if game is possible.

2 When you rebid your own suit, or raise your partner's, the
   higher you bid, the stronger you are. The same is true when
   you rebid in no-trumps. All such rebids are limit bids.

3 A rebid in a new suit (such as 1♥ – 1♠ – 2♣) is wide-range.
   You may have a minimum opening hand; you may instead
   have a middle-range hand. A jump rebid in a new suit (1♥ –
   1♠ – 3♣) shows a very strong hand and is game-forcing.

4 A rebid at the two level in a suit higher than your first suit (such
   as 1♣ – 1♠ – 2♥) is known as a 'reverse'. Because it pushes the
   bidding so high, you need around 17 points to reverse.

Quiz

a After a start of 1♣ – 1♠, which rebids by the opener would
   be limit bids?

b What is meant by a 'reverse'?

c In the sequence 1♦ – 1♠ – 3♦, is the opener's rebid forcing?

d In the sequence 1♥ – 2♣ – 3♥, is the opener's rebid forcing?

e In the sequence 1♦ – 1♠ – 2♣, what range of points may the
   opener have?

Answers

a All rebids in clubs, spades, or no-trumps would be limit bids.
   Only a rebid in a new suit would be wide-range.

b A reverse is a rebid at the two level in a suit higher than that
   of the opening bid (for example, 1♣ – 1♠ – 2♥).

c No, it is a limit bid, showing a hand in the middle range.

d Yes, when the response is at the two level (showing at least
   10 points) a jump rebid is forcing to game.

e A rebid in a new suit is wide-range. The opener may have
   anything from a minimum hand of around 12 points to a full
   17 points.

8 THE RESPONDER'S REBID

By the time the responder comes to make his second bid, it will often be possible to determine a trump suit (if any) and the level at which the contract should be played. In the majority of auctions one or other player will have made a limit bid by now; his partner can then calculate the final contract.

There are many possible situations and we will cover them in this order:

■ the opener made a simple rebid in his suit,
■ the opener made a jump rebid in his suit,
■ the opener rebid in no-trumps,
■ the opener raised the responder's suit,
■ the opener introduced a new suit.

The opener made a simple rebid in his suit

When the opener rebids his suit at the minimum level he suggests a hand of around 12 to 14 points. The responder can now judge whether to try for game, depending on his own strength:

Weak	6–9 points	Play in a part score
Medium	10–12 points	Make a game try
Strong	13+	Head for game

As you will realise by now, it is not only a question of points. A long suit, or good support for your partner, will add considerably to the value of your hand.

After a start of, say, 1♥ – 1♠ – 2♥, these are the limit bids available to you:

Weak bids:	Pass	2♠	
Game tries:	2NT	3♥	3♠
Game bids:	3NT	4♥	4♠

Suppose the bidding has indeed started 1♥ – 1♠ – 2♥ and, as responder, you hold one of these hands in the 'weak' range:

(1)   ♠ A 8 4 3   (2)   ♠ A Q 10 8 7 2
       ♥ 2   ♥ 8
       ♦ 10 7 3   ♦ Q 4 2
       ♣ K J 9 7 4   ♣ 9 4 3

There is no chance of game on hand (1) and you should pass. Do not make the mistake of bidding further because you 'do not like hearts'. The occasions when you have no trump fit with your partner are exactly those when you should keep the contract as low as possible. On hand (2) your spades are good and there is a fair chance that spades will make a better trump suit than hearts. Bid 2♠, again showing a hand in the weak range.

Now let's move to some hands in the game-try range of 10–12 points. Again we will assume the bidding has started 1♥ – 1♠ – 2♥.

(3)   ♠ A Q 6 2   (4)   ♠ K J 6 5 2   (5)   ♠ A Q 9 8 7 3
       ♥ Q 8 4   ♥ 8   ♥ 5
       ♦ 10 2   ♦ Q 10 5   ♦ K Q 6
       ♣ K 9 6 5   ♣ A J 8 3   ♣ 10 6 2

On hand (3) you raise to 3♥, suggesting game in hearts. Your partner's rebid in hearts promises at least five cards there, so you know that you have a satisfactory trump fit of at least eight cards.

There is no trump fit on hand (4) but your 11 points justify a game try. You would bid 2NT, showing that you have both the unbid suits stopped. Your partner may then pass on a minimum hand. With anything extra, he will raise to 3NT (or perhaps bid a forcing 3♠ to show three-card support).

On hand (5) your game-try bid would be 3♠, showing around 11 points and a good spade suit of at least six cards. Your partner would then have the option of passing, or of bidding game somewhere.

When you hold 13 points, or slightly fewer points but a very good suit or fit, you must carry the bidding to game. On these two hands (again after a start of 1♥ – 1♠ – 2♥) you would know already which game to bid:

(6)   ♠ K J 7 5 2   (7)   ♠ A Q 9 7
       ♥ A J 3   ♥ 8 2
       ♦ 7 3   ♦ A J 4
       ♣ A 10 3   ♣ K 10 6 2

On (6) you know of an eight-card heart fit, at least, and should raise to 4♥. On hand (7) you would bid 3NT.

When you have a hand strong enough for game but do not yet know which game will be best, you can bid a new suit.

(8) ♠ A Q 10 7 6
     ♥ J 3
     ♦ 7 3
     ♣ A Q 6 2

After 1♥ – 1♠ – 2♥, you would continue with 3♣. This shows a strong hand, good enough for game somewhere. If your partner has a stopper in diamonds he may now be able to bid 3NT. He may instead bid his hearts again; you would then place him with six hearts and bid game in that suit. Finally he might bid 3♠, suggesting three-card spade support (with four he would have raised spades on the previous round). You would then bid game in spades.

The opener made a jump rebid in his suit

When the opener makes a jump rebid such as 1♥ – 1♠ – 3♥, he suggests around 15–17 points and a good six-card suit. You may pass if you hold a near minimum response and no fit for his suit. If instead you make any bid at all, the bidding must continue to game.

After a start of 1♥ – 1♠ – 3♥ what would you say next on these hands:

(1) ♠ K J 7 5   (2) ♠ 10 8 7 2
     ♥ 3   ♥ K 2
     ♦ J 5 3   ♦ 9 6
     ♣ Q 10 6 5 3   ♣ A 10 6 4 2

Hand (1) is a near-minimum response of seven points, with no fit for hearts. You would pass. Hand (2) also contains seven points, but you have a useful king-doubleton in trumps, a precious ace, and a possible ruffing value in diamonds (your partner may be able to make extra tricks by ruffing diamonds in your hand). You would raise to 4♥.

The opener rebid in no-trumps

When your partner has rebid in no-trumps, you add your own points to those he has shown. You will then know whether to play in a part score, try for game, or bid game.

Suppose the bidding has started 1♦ – 1♠ – 1NT. These limit bids are available to you:

Weak rebids	Pass	2♦	2♠
Game-try rebids	2NT	3♦	3♠
Game bids	3NT	4♠	and (rarely) 5♦

The actual strength needed for a game try or a game bid will depend on whether the 1NT rebid is 15–17 (UK) or 12–14 (USA).

When you hold a strong hand and cannot yet decide on the best game, you can bid a new suit at the three level.

(1) ♠ A K 10 7 6
♥ 3
♦ Q 7
♣ K Q 10 6 2

After 1♦ – 1♠ – 1NT you would jump to 3♣, forcing to game. Your partner may be able to support the spades or the clubs. Failing that, you can play in 3NT.

When your partner makes a jump rebid of 2NT (as in 1♦ – 1♠ – 2NT) he suggests 18–19 points. You are allowed to pass if you responded on a poor hand of only six points. If instead you make any bid, including 3♦ or 3♠, this is forcing to game.

The opener raised the responder's suit

When the opener gives a single raise of your suit he will be in the minimum range. You will need around 10–12 points to make a game try, 13 points to bid game. Suppose the bidding has started 1♦ – 1♥ – 2♥ and you hold one of these hands:

(1)	♠ K 10 6 2	(2)	♠ 7 4	(3)	♠ A 5
	♥ A J 6 3		♥ A 10 9 7 4		♥ A J 8 6
	♦ 10 2		♦ Q 10 5		♦ K J 6
	♣ J 8 3		♣ A 9 3		♣ 10 9 6 2

On (1) you have only nine points. Game is unlikely and you should pass. Hand (2) is appreciably stronger. With ten points and an important fifth trump, you can invite game by bidding 3♥. Your partner can then choose between passing and advancing to game. Hand (3) has 13 points, enough for game facing an opening bid. You would bid 4♥ now.

When the opener gives a double raise of your suit he will be in the middle range, 15–17 points. You will need around nine points to bid game.

The opener introduced a new suit

In all the situations so far, the opener's rebid gave a clear picture of his strength. It was therefore possible for the responder to 'announce the answer'. The situation is different when the opener introduces a new suit (a sequence such as 1♥ – 1♠ – 2♣). The rebid of 2♣ is wide-range. The opener may hold a shapely ten-count; he may hold as many as 17 points.

Because the range is so wide the responder may not be in a position to determine the final contract. In this case he must instead give a picture of his own strength, allowing the opener to judge the game prospects.

After a start of 1♥ – 1♠ – 2♣ the responder has these limit bids:

Weak	6-9 points	Pass	2♥ 2♠
Game try	10-12 points	2NT	3♣ 3♥ 3♠
Game bids	13+ points	3NT	4♥ 4♠ and (rarely) 5♣

There is one important situation that you will meet time and time again. The bidding has started 1♥ – 1♠ – 2♣ and you hold one of these hands:

(1)	♠ A J 6 3	(2)	♠ A J 6 3
	♥ 10 4 3		♥ 9 2
	♦ Q 7 6 2		♦ J 10 8 6 3
	♣ J 5		♣ Q 8

Both hands are in the 6–9 weak range and, as far as you can judge, you want to play in a part score. There is no problem whatsoever on (1). You expect your partner to hold five hearts and four (possibly five) clubs. Knowing that hearts will make a good trump suit, you are happy to bid 2♥. Such a bid is known as 'giving preference'. Your partner has shown two suits and you are merely showing which of them you prefer. Bidding 2♥ does not show any more values than a pass of 2♣ would have done.

Now look at hand (2). Again you are in the weak range. Which of your partner's suits do you prefer? The answer is hearts. You have equal length in both his suits, but your partner's hearts are likely to be longer than his clubs. On most hands he will hold five hearts and only four clubs. So, strange as it may seem, you should bid 2♥ now. You are not showing support, you are giving preference. Your partner will realise that you may have been forced to make the bid with only two-card support.

There is usually no problem when your hand is in the game-try range. Again assume a start of 1♥ – 1♠ – 2♣.

(3)   ♠ K 10 6 2   (4)   ♠ A J 8 6 3   (5)   ♠ A Q 9 8 5 2
        ♥ 4 2   ♥ 10 9 4   ♥ J
        ♦ K J 6 3   ♦ A J 8 5   ♦ K 7 6
        ♣ A 8 2   ♣ 5   ♣ Q 9 6

You bid 2NT on (3), suggesting game in no-trumps; your partner may now pass if he has a minimum hand. On hand (4) you would bid 3♥, inviting a heart game. This is known as 'jump preference'. On (5) you would choose 3♠, showing a six-card spade suit and sufficient values to invite game.

Make the hands slightly stronger and you will be able to bid game yourself.

(6)   ♠ K 10 6 2   (7)   ♠ A J 8 6 3   (8)   ♠ A K Q 8 5 2
        ♥ 4 2   ♥ K 9 4   ♥ J
        ♦ A Q 6 3   ♦ A J 8 5   ♦ K 7 6
        ♣ A 8 2   ♣ 5   ♣ Q 9 6

After a start of 1♥ – 1♠ – 2♣, you would bid 3NT on (6), 4♥ on (7), and 4♠ on (8).

The 'fourth suit forcing' convention

As we saw in the previous section a responder with game values can usually choose which game to bid after a start such as 1♥ – 1♠ – 2♣. Sometimes, though, he will find himself confronted with a hand like this:

                        ♠ A K J 7 6
                        ♥ 6 3
                        ♦ 9 7 6
                        ♣ A Q 4

Enough for game, yes, but which game? If his partner has some sort of spade support, 4♠ may be best. If his partner has a diamond stop, 3NT may be a good contract. Unfortunately, there seems to be no forcing bid available, to keep the auction heading towards game. Bids such as 3♣ and 3♠ would both be non-forcing, showing a hand in the 10–12 range.

Or maybe you hold this, equally awkward, hand:

♠ A K J 7
♥ 6 3
♦ 9 7 6
♣ A Q 8 5

Enough for game again, and this time we hold four-card support for our partner's club suit. But 3♣ would be non-forcing and 4♣ would carry us past 3NT, which is where we want to play if our partner holds a diamond stopper.

What we need is some bid which will pass the message 'I have enough for game, partner, but no idea at the moment which game will be best.'

Such a bid is available! You bid the 'fourth suit', the only suit which has not yet been bid:

West	East
1♥	1♠
2♣	2♦

Such a bid has a universal meaning, wherever bridge is played. It says nothing about your holding in the suit actually bid (here diamonds); it merely tells your partner that you hold a hand strong enough for game.

On both the hands we saw, you would bid 2♦, the fourth suit. The bidding will then continue towards game. Your partner will have the opportunity to bid no-trumps, if he holds a diamond stopper. He may also be able to rebid his second suit (to show a 5–5 hand) or to show delayed support for your own suit.

Let's see how 'fourth suit forcing' works on some auctions that involve the two hands above.

West	East	West	East
♠ 8	♠ A K J 7 6	1♥	1♠
♥ A J 9 8 2	♥ 6 3	2♣	2♦
♦ A 10 3	♦ 9 7 6	2NT	3NT
♣ K J 5 2	♣ A Q 4	End	

East bids the fourth suit (2♦) to show that he has enough for game. West now bids 2NT, indicating a stopper in diamonds, the unbid suit. With this knowledge East can bid game in no-trumps.

West	East	West	East
♠ Q 5	♠ A K J 7 6	1♥	1♠
♥ A K J 9 2	♥ 6 3	2♣	2♦
♦ 10 3	♦ 9 7 6	2♠	4♠
♣ K 10 5 2	♣ A Q 4	End	

Here West chooses to bid 2♠, showing some sort of spade support. (If he held three spades he would probably have jumped to 3♠ instead). East now bids game in spades. This is a better spot than 3NT since the defenders might be able to score five diamond tricks in no-trumps.

West	East	West	East
♠ 8	♠ A K J 7	1♥	1♠
♥ A J 10 9 4 2	♥ 6 3	2♣	2♦
♦ Q 3	♦ 9 7 6	2♥	3♣
♣ K J 4 2	♣ A Q 8 5	3♥	4♥
		End	

Over the fourth-suit call West bids his hearts again. East can now show his club support (3♣ is forcing because it was preceded by a fourth-suit call). West bids his hearts yet again, to show a six-card suit, and East is happy to play game in hearts.

West	East	West	East
♠ 8	♠ A K J 7	1♥	1♠
♥ A K J 9 2	♥ 6 3	2♣	2♦
♦ 4 3	♦ 9 7 6	3♣	5♣
♣ K J 10 4 2	♣ A Q 8 5	End	

This time West rebids his second suit, to show five cards there. East decides to play in the club game.

Points to remember

1   When the opener has made a limit bid, the responder adds his own values to his partner's and announces the answer: part score, game try, or game bid. In general, the higher he bids, the stronger the hand he holds.

2   When the opener has shown two suits (a sequence such as 1♥ – 1♠ – 2♣), the responder may have to 'give preference' to the first suit when he holds only a doubleton there.

3 When the responder has enough for game but does not yet know which game will be best, he may use the 'fourth suit forcing' convention.

Quiz

a After a start of 1♦ – 1♥ – 2♦, which calls by the responder would show a hand in the minimum range (6–9 points)?

b After a start of 1♦ – 1♠ – 2♣, which calls by the responder would show a minimum hand?

c What does it mean to 'show preference'?

d After a start of 1♥ – 2♣ – 2♦, which bid by the responder would be 'fourth suit forcing'?

Answers

a Pass or 2♥.

b Pass, 2♦ or 2♠.

c To choose one of the suits that your partner has bid.

d 2♠.

9 | TECHNIQUES IN NO-TRUMPS

Let's take a break from bidding and look at some of the basic cardplay techniques available to the declarer, first of all in no-trump contracts. The most important of these is known as the 'hold-up'.

When you have a high card such as an ace in the suit the opponents have led, it is not always right to win a trick with it at the first opportunity. Suppose you are playing in 3NT and the spade suit lies like this:

```
                    North
                    ♠ 8 3
     West                          East
♠ K Q J 10 2                       ♠ 9 7 4
                    South
                    ♠ A 6 5
```

West leads ♠K (top of a sequence). Suppose you win the first trick with the ace. If East gains the lead later in the play, he will be able to return a spade; West will score four spade tricks to defeat the contract.

Instead you should refuse to win the ace until the third round of the suit. East will then have no spades left. He will not be able to return a spade, should he gain the lead later.

The same would be true here:

```
                    North
                    ♠ 10 3
     West                          East
♠ Q J 7 5 2                        ♠ A 9 4
                    South
                    ♠ K 8 6
```

West leads ♠5 to East's ace and back comes ♠9. If you take your king immediately, East will still have a spade left. Hold up the king for one more round and you exhaust East's spades.

Let's see how this works in the context of a whole deal. Suppose you are playing 3NT here:

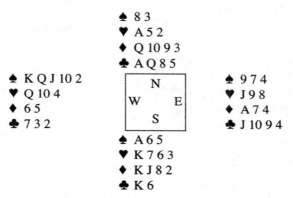

```
                    ♠ 8 3
                    ♥ A 5 2
                    ♦ Q 10 9 3
                    ♣ A Q 8 5
  ♠ K Q J 10 2        ┌─────────┐      ♠ 9 7 4
  ♥ Q 10 4            │    N    │      ♥ J 9 8
  ♦ 6 5              │ W     E │      ♦ A 7 4
  ♣ 7 3 2            │    S    │      ♣ J 10 9 4
                     └─────────┘
                    ♠ A 6 5
                    ♥ K 7 6 3
                    ♦ K J 8 2
                    ♣ K 6
```

Trick 1 West leads ♠K against 3NT. You hold up the ace, allowing the king to win.

Trick 2 West continues with ♠Q. Again you hold up the ace.

Trick 3 West now plays ♠J. A heart is thrown from dummy, East follows with his last spade, and you win with the ace.

At this stage you can count only six certain tricks (one spade, two hearts, and three clubs). Three more tricks can be obtained from the diamond suit, by knocking out the defenders' ace.

Trick 4 You lead a diamond to the queen.

If East refuses to take the ace of diamonds you will continue to play diamonds until he does. The contract will succeed because, thanks to your hold-up, East will have no spade to play when he gains the lead. You can win whatever other suit he returns and score the nine tricks you need.

'The hold-up wouldn't have worked if West held the ace of diamonds,' you may be saying. Quite right, but in that case the contract could not be made at all. By holding up the spade ace twice you did at least give yourself the best chance of making the contract.

Playing into the safe hand

After a hold-up, one defender may be 'safe' (he has no cards left in the defenders' long suit), the other may be 'dangerous'. When you are

deciding how to establish the tricks you need to make the contract, you
should choose a line of play which will allow only the safe defender to
gain the lead. Look at this deal:

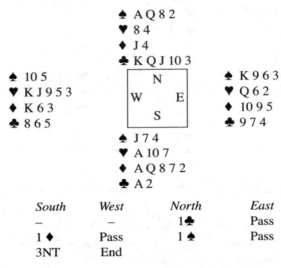

♠ A Q 8 2
♥ 8 4
♦ J 4
♣ K Q J 10 3

♠ 10 5
♥ K J 9 5 3
♦ K 6 3
♣ 8 6 5

♠ K 9 6 3
♥ Q 6 2
♦ 10 9 5
♣ 9 7 4

♠ J 7 4
♥ A 10 7
♦ A Q 8 7 2
♣ A 2

South	West	North	East
–	–	1 ♣	Pass
1 ♦	Pass	1 ♠	Pass
3NT	End		

West leads ♥5 (his fourth best heart) against 3NT. You hold up the ace
until the third round, exhausting East of hearts. You can now count eight
certain tricks: five clubs and three other aces. A successful spade finesse
would bring the total to nine; so would a successful finesse in diamonds.
Which finesse should you take, do you think?

Suppose you take a diamond finesse and West wins with the king. The
contract will now go down because West will be able to make two more
tricks with his long hearts. The diamond finesse was 'into the danger hand'.

Suppose instead that you follow the wiser course of finessing in spades.
Dummy's ♠Q will lose to East's king but you will still make the contract!
East has no heart to play, after your hold-up, and when you regain the lead
you can make an extra trick with your ♠J. The spade finesse was 'into the
safe hand' and that is why it was the better play to choose.

The next deal is similar. See if you can work out which defender has
become 'dangerous' and therefore how you should play the hand.

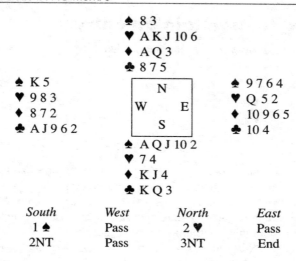

	♠ 8 3	
	♥ A K J 10 6	
	♦ A Q 3	
	♣ 8 7 5	

South	West	North	East
1 ♠	Pass	2 ♥	Pass
2NT	Pass	3NT	End

West leads ♣6 against 3NT. East plays the 10 and you win the first trick with the queen. How will you play the hand?

The first thing to do, as always, is to count the top tricks. You have one spade, two hearts, three diamonds, and one club – a total of seven. If West held ♥Q you could finesse dummy's ♥J, return to the South hand with a diamond, and finesse ♥10. You would probably score ten tricks, more than enough for the contract. The same is true if East held ♠K. You could take two spade finesses and make extra tricks there.

Which finesse do you think you should take? The point to consider is: what will happen if the chosen finesse fails? If a heart finesse loses, East will gain the lead and disaster will ensue. He will lead a club through your king and West will then score four club tricks to beat the contract. East is the 'danger hand' here, because he is the one who can lead through your ♣K.

What will happen if instead you take a spade finesse and it loses? The answer is that you will still make the contract. West cannot play on clubs from his side of the table without allowing you to make a trick with ♣K. Once again the correct way to play the hand was to take the finesse that was 'into the safe hand'.

Ducking to maintain an entry

Suppose you need tricks from a suit that lies like this:

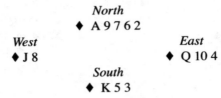

The ace and king are certain to score tricks, but you would like to make some tricks from the small cards as well (as we saw in Chapter 2). One way to do this would be to make a trick with ♦K, then a trick with ♦A. You would then play a third round of diamonds, allowing East's queen to win. The last two diamonds in dummy would then be 'good'. No other diamonds would still be out.

An equally good way to create four tricks would be to 'duck' (in other words, allow the opponents to win) the first or second round of the suit. For example, you could make a trick with ♦K, then play a low diamond from your own hand and the dummy, letting the defenders win. This would often be a better way to play the suit because the ace of diamonds would remain in dummy as an entry.

Suppose this is the full hand:

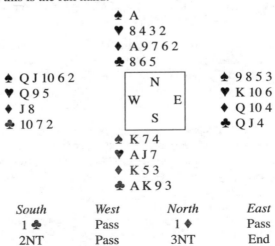

South	West	North	East
South	*West*	*North*	*East*
1 ♣	Pass	1 ♦	Pass
2NT	Pass	3NT	End

West leads ♣Q, won with the dummy's ace. You can count seven top tricks (two spades, one heart, two diamonds and two clubs). Two extra tricks in diamonds would bring the total to nine. However, look what would happen if you next made tricks with the king and ace of diamonds. You could give up the third round of diamonds to East, but on regaining the lead you would have no way to reach the two good diamonds in the dummy. The contract would go down.

You need ♦A as an entry to dummy. After winning the spade lead with the ace, you should play a diamond to the king, then duck a round of diamonds. Everything is easy after that. You can win the spade return and cross to dummy with ♦A to score tricks with the last two diamonds. That will give you nine tricks in all.

Sometimes you will need to duck two rounds of a suit before you can enjoy tricks with the small cards. You may hold a suit like this:

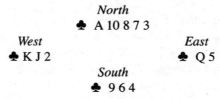

North
♣ A 10 8 7 3

West　　　　　　　　　　　　*East*
♣ K J 2　　　　　　　　　　　♣ Q 5

South
♣ 9 6 4

Duck the first two rounds of the suit. You will then be able to cross to ♣A and make two tricks with dummy's last two clubs.

This might be the full hand:

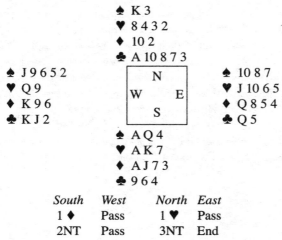

♠ K 3
♥ 8 4 3 2
♦ 10 2
♣ A 10 8 7 3

♠ J 9 6 5 2　　　　　　　　　　♠ 10 8 7
♥ Q 9　　　　　　　　　　　　♥ J 10 6 5
♦ K 9 6　　　　　　　　　　　♦ Q 8 5 4
♣ K J 2　　　　　　　　　　　♣ Q 5

♠ A Q 4
♥ A K 7
♦ A J 7 3
♣ 9 6 4

South	West	North	East
1 ♦	Pass	1 ♥	Pass
2NT	Pass	3NT	End

West leads ♠5 against 3NT. Can you see how the play will go, trick by trick, for you to make nine tricks? It should go like this:

Trick 1 You win the spade lead with the queen.

Trick 2 You duck a club (you play a low club from both hands), East winning with the queen.

Trick 3 East returns a spade, won with dummy's king.

Trick 4 You now duck another club, West winning with the jack.

Trick 5 West plays a third round of spades, won with your ace.

Trick 6 You reach dummy with the carefully preserved ♣A,

Trick 7 ... make a trick with dummy's penultimate club

Trick 8 ... and make a trick with dummy's last club.

Tricks 9–11 You score further tricks with ♥A K and ♦A, bringing your total to nine.

Tricks 12–13 The defenders score the last two tricks.

You see how important it was to duck the first two rounds of clubs? The point was that the defenders would remove dummy's ♠K as soon as they regained the lead. Since your clubs would not be established by then, it was clear that he would need the ace of clubs as an entry to reach the long cards in clubs.

Points to remember

1   Declarer's most important technique in a no-trump contract is the hold-up. By refusing to win an ace until one of the defenders is exhausted in the suit, you can often avoid the eventual loss of tricks to the long cards in the defenders' suit.

2   It often happens that one defender is 'dangerous' (if he gains the lead he will have tricks to cash); the other defender is 'safe'. You must then choose a line of play which will not allow the dangerous defender to gain the lead.

3   To establish a suit, it may be necessary to give up a trick or two in the suit. By ducking an early round (or two) you can retain a high card as a later entry.

Quiz

a

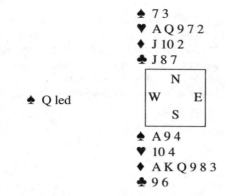

♠ 7 3
♥ A Q 9 7 2
♦ J 10 2
♣ J 8 7

♠ Q led

♠ A 9 4
♥ 10 4
♦ A K Q 9 8 3
♣ 9 6

West leads ♠Q against 3NT. How will you plan the play? In particular, for how many rounds will you hold up your ♠A. (Don't answer too quickly. Make sure you look at the whole hand.)

b

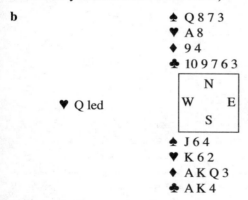

♠ Q 8 7 3
♥ A 8
♦ 9 4
♣ 10 9 7 6 3

♥ Q led

♠ J 6 4
♥ K 6 2
♦ A K Q 3
♣ A K 4

West leads ♥Q against 3NT. Which nine tricks will you aim to make? How will you plan the play?

Answers

    a You cannot afford to hold up ♠A at all! If you do, the opponents may immediately take at least four club tricks. Win the first spade trick and finesse ♥Q. If the finesse succeeds you will have nine tricks (six diamonds, two hearts, and one spade).

b You plan to make four clubs, three diamonds and two hearts. Win the heart lead with the king, retaining ♥A as a later entry to dummy. Now play the ace and king of clubs. If the suit breaks 3–2 (or if a singleton queen or jack falls) you can clear the club suit by playing a third round. When you regain the lead you can use ♥A as the entry to the two club winners. You will have nine tricks.

10 | TECHNIQUES IN A SUIT CONTRACT

The presence of a trump suit gives you more options as declarer. You can ruff losing cards in the dummy; you can prevent the defenders from scoring tricks by ruffing their winners. You can also establish a side suit by taking one or more ruffs. First we will see how it is possible to make extra tricks by ruffing.

Making extra tricks by ruffing

Suppose you are playing in a contract such as Four Spades and this is the trump suit:

North
♠ Q 4 3

West   *East*
♠ 8 5   ♠ J 10 2

South
♠ A K 9 7 6

How many trump tricks will you make, altogether, if you ruff one side-suit card in the dummy? The answer is six. You will make five trump tricks in your hand, plus one ruff in the dummy.

How many trump tricks will you make if instead you ruff one side-suit card in your own hand? You will not do so well; you will make only the five trump tricks that you started with. It's an important idea to grasp. Ruffing in the shorter trump holding will give you an extra trick. Ruffing in the longer trump holding will not give you an extra trick.

Look at this complete deal, which involves the trump holding above:

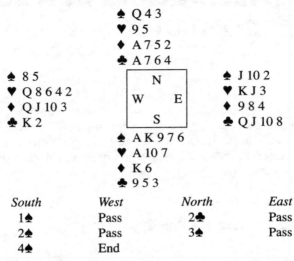

South	*West*	*North*	*East*
1♠	Pass	2♣	Pass
2♠	Pass	3♠	Pass
4♠	End		

West leads ♦ Q against Four Spades and you win with the king. If the opposing trumps break 3–2, you have nine top tricks: five trumps, three aces and ♦ K. Suppose you cash the other top diamond and ruff a diamond in your own hand. Will that help at all? No, you will still have only nine tricks.

To bump the total to ten you need to ruff something in the dummy, taking a ruff in the shorter trump holding. Which suit can you ruff in the dummy? You can ruff a heart, because dummy's hearts are shorter than your own.

This is how the play would go:

Trick 1 You win the diamond lead with the king,

Trick 2 ... and lead a low heart from hand (since a heart trick must be given up before taking the ruff). East wins with the jack.

Trick 3 East returns ♣ Q, won with the ace.

Trick 4 You play a heart to the ace,

Trick 5 ... and ruff a heart.

Tricks 6–8 You now draw trumps in three rounds, the queen first, then the ace and the king.

Tricks 9–11 You play a diamond to the ace and score the next two tricks with your last two trumps.

Tricks 12–13 The defenders take the remaining two tricks.

You had nine tricks initially, provided the outstanding trumps broke 3–2. You made a tenth trick by ruffing a heart in the short trump holding.

Ruffing high to avoid an overruff

Whenever possible you should ruff with a high trump, rather than a low trump. This will prevent the next player from overruffing, should he also be out of the suit that you led.

On the previous hand dummy's trumps were Q 4 3 and you held A K 9 7 6. You could not afford to ruff with dummy's queen, because the defenders would then have scored a trump trick.

The situation is different on this deal:

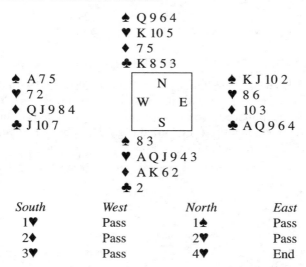

```
                    ♠ Q 9 6 4
                    ♥ K 10 5
                    ♦ 7 5
                    ♣ K 8 5 3
   ♠ A 7 5                              ♠ K J 10 2
   ♥ 7 2          ┌─────────┐          ♥ 8 6
   ♦ Q J 9 8 4    │   N     │          ♦ 10 3
   ♣ J 10 7       │ W     E │          ♣ A Q 9 6 4
                  │   S     │
                  └─────────┘
                    ♠ 8 3
                    ♥ A Q J 9 4 3
                    ♦ A K 6 2
                    ♣ 2
```

South	West	North	East
1♥	Pass	1♠	Pass
2♦	Pass	2♥	Pass
3♥	Pass	4♥	End

When North gives preference to hearts, South invites game by raising to 3♥. North is short in his partner's second suit (diamonds) and can therefore see that extra tricks may be available by ruffing diamonds in the shorter trump holding. He advances to the heart game.

West leads ♣ J, winning the first trick. He continues with ♣ 10 and you ruff in the South hand. There are eight top tricks: six trumps and the ace-king of diamonds. You can score the two extra tricks you need by ruffing diamonds in dummy. Suppose you play the ace and king of diamonds and ruff a diamond with dummy's 5 of trumps. Disaster! East started with only two diamonds and will overruff with the 6. The contract will go one down.

To avoid this fate, you must ruff the third round of diamonds with the 10. East cannot overruff and you return to your hand by leading the 5 of trumps to the ace. Now you ruff your last diamond with the king. You cross back to your hand by ruffing a club (you can afford to take this ruff with a high trump too, although in fact West still has a club left). You then draw trumps, making ten tricks and your contract.

Establishing a suit by ruffing

Suppose that spades are trumps and you have a side suit distributed like this:

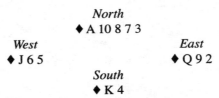

```
                    North
                    ♦ A 10 8 7 3
        West                        East
        ♦ J 6 5                     ♦ Q 9 2
                    South
                    ♦ K 4
```

You play ♦ K and cross to dummy with ♦ A. You then lead a third round of diamonds from dummy and ruff in your own hand. Because the suit broke 3–3 the opponents now have no more diamonds. You have established the diamond suit (or 'ruffed it good', as bridge players say). The last two diamonds in dummy will score tricks for you.

Let's see how you might use this technique in the context of a full deal.

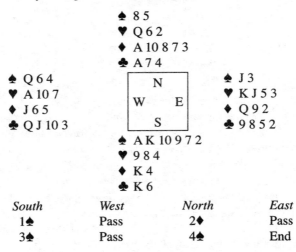

```
                    ♠ 8 5
                    ♥ Q 6 2
                    ♦ A 10 8 7 3
                    ♣ A 7 4
    ♠ Q 6 4                         ♠ J 3
    ♥ A 10 7          N             ♥ K J 5 3
    ♦ J 6 5       W       E         ♦ Q 9 2
    ♣ Q J 10 3       S             ♣ 9 8 5 2
                    ♠ A K 10 9 7 2
                    ♥ 9 8 4
                    ♦ K 4
                    ♣ K 6
```

South	West	North	East
1♠	Pass	2♦	Pass
3♠	Pass	4♠	End

A heart lead would have beaten the contract, as it happens. (The defenders would have taken three heart tricks and eventually made a trump trick.) West in fact made the normal 'top of a sequence' lead, the queen of clubs. How would you have played the contract after this lead?

You should win the club lead with the king, retaining ♣ A as a later entry to the dummy. You then draw two rounds of trumps with the ace and king. There is one trick to be lost in trumps and you must somehow avoid three more losing tricks in hearts. The best chance is to ruff the diamonds good. You will need some luck for this to be possible; the opposing diamonds will have to break 3–3.

You cash ♦ K, cross to ♦ A, and ruff a diamond. Luck is with you and both defenders follow suit all the way. You now cross to dummy's ♣ A and lead one of the good diamonds, throwing a losing heart. West ruffs with the queen of trumps but the defenders can score only this card and two heart tricks. You will take the remaining ten tricks, making the contract.

Sometimes you need to ruff a suit good because the defenders' cards have not broken evenly. You might have this diamond side suit:

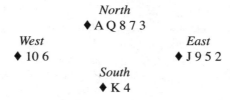

```
                North
              ♦ A Q 8 7 3
  West                        East
  ♦ 10 6                      ♦ J 9 5 2
                South
              ♦ K 4
```

If the defenders' cards were divided 3–3, you would make five diamond tricks. As the cards lie here, West would show out on the third round of diamonds. You would then ruff the fourth round of the suit, setting up dummy's last diamond.

Planning a trump contract

How do you plan the play when you are in a trump contract? Until now we have been using the same method as in no-trumps, counting the top tricks and then deciding how to create the extra tricks we needed.

There is another way that many people find easier when playing in a trump contract. You count the 'losers' (tricks which you might lose) in each of the four suits. You then see how you can reduce this number of losers to the required total.

It's not at all difficult, as you will see here.

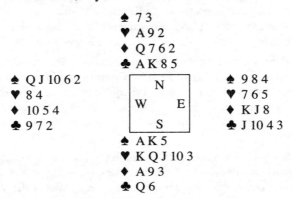

```
                        ♠ 7 3
                        ♥ A 9 2
                        ♦ Q 7 6 2
                        ♣ A K 8 5
  ♠ Q J 10 6 2          ┌─────────┐          ♠ 9 8 4
  ♥ 8 4                 │    N    │          ♥ 7 6 5
  ♦ 10 5 4              │ W     E │          ♦ K J 8
  ♣ 9 7 2               │    S    │          ♣ J 10 4 3
                        └─────────┘
                        ♠ A K 5
                        ♥ K Q J 10 3
                        ♦ A 9 3
                        ♣ Q 6
```

Let's suppose you are in Six Hearts, a contract where you can afford only one loser. West leads ♠ Q and you win with the ace. To plan the contract, you count the possible losers in the hand with the long trumps (here South), taking each suit in turn:

Spades	1 loser	♠ 5
Trumps	0 losers	(you have all the top trumps)
Diamonds	2 losers	♦ 9 3
Clubs	0 losers	(you have all the top clubs)

So, there are three possible losers. You need to reduce them to one, to make the slam.

Is there any way to avoid losing a trick with ♠ 5? Yes, you can ruff it in the dummy. The only remaining problem is to reduce the two diamond losers to just one. Can you see how to do this?

One possibility would be to lead a low diamond towards dummy's queen. If you were lucky and West held ♦ K, you would then lose only one diamond. But there is a certain way to avoid losing two diamonds; you can throw one of them away on the third round of clubs.

The play will go like this. You win the spade lead with the ace, play ♠ K and ruff a spade with the 9 of trumps (ruffing high, in case East could otherwise overruff). You then draw trumps, play ♣ Q, play a club to the ace and discard one of your diamonds on ♣ K.

Points to remember

1 To plan a suit contract, you count the possible losers in the hand with the long trumps. In Four Spades, for example, you can afford three losers. If you start with five losers you will have to plan how you can reduce them to three.

2 The three main ways of avoiding possible losers are: finessing, ruffing losers in the dummy, or discarding losers on winners in the dummy.

3 When taking ruffs, ruff with high trumps whenever you can afford to do so. This will prevent the defenders from overruffing.

4 You can often establish a side suit by ruffing it until the defenders have no cards left in the suit.

Quiz

a Is it better to take ruffs in the shorter trump holding or the longer trump holding?

b You have a side suit of ♣ A K 8 5 3 (dummy) and ♣ 9 4 (your hand). To establish the suit with only one ruff, how will the defenders' clubs have to break? How will their clubs have to break for you to establish the suit with two ruffs?

c Why might you sometimes choose to ruff with a high trump rather than a low one?

d Study the hand overleaf. West leads ♥ Q against Six Spades. Count the possible losers in the long-trump hand (South). How many losers are there in each of the four suits? What is the total number of possible losers? How do you plan to reduce the number to just one?

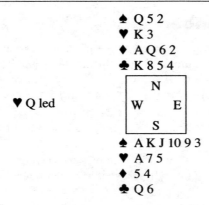

♠ Q 5 2
♥ K 3
♦ A Q 6 2
♣ K 8 5 4

♥ Q led

♠ A K J 10 9 3
♥ A 7 5
♦ 5 4
♣ Q 6

Answers

a It is better to take ruffs in the shorter trump holding. Each ruff will then produce an extra trick.

b If the defenders' clubs break 3–3 you can set up the suit with only one ruff. If they break 4–2 you will need to take two ruffs.

c To remove the risk of a defender overruffing the low trump.

d There are no losers in trumps, one possible heart loser, one possible diamond loser, and one certain club loser (you will lose to the ace). That is a total of three possible losers. You can avoid the heart loser by ruffing it in dummy. If West holds ♦ K you can avoid the diamond loser by taking a successful finesse.

11 THE SCORING

There is more than one variant of bridge, as we will see in Chapter 19. In the traditional form, 'rubber bridge', the side that is first to score two games wins the rubber. The scores are then calculated and, if there was a stake involved, the amount is settled.

Each player has a scoresheet, which looks like this:

We	They

Any points scored by your side will be written on the left side of the sheet, those scored by your opponents on the right side. Points for tricks bid and made are written 'below the line' and count towards the 100 points needed for a game. As we saw in Chapter 1, these are the trick values:

Clubs or diamonds are trumps	20 per trick
Hearts or spades are trumps	30 per trick
No-trumps	40 for 1st trick, then 30 per trick

Let's suppose that on the first deal of the rubber your side bids Two Spades and makes nine tricks. You score 60 below the line, for the two tricks bid and made. For the overtrick, you score 30 above the line. This will contribute to your final total but will not count towards game. The scoresheet will now look like this:

We		They
30		
60		

If you could score 40 or more on the next deal, you would bring your total below the line to 100 and win the first game. Unfortunately, in our imaginary rubber, the opponents now pick up some good cards. They bid 3NT and make it exactly. They score 100 below the line, giving them the first game. A line is drawn under the scores so far, indicating that the first game is over. The sheet now looks like this:

We		They
30		
60		100

Your 60 points for a part score were part of the first game and will not contribute further, except in the final addition. Both sides must now strive to achieve 100 points afresh.

Leaving this rubber for a moment, we will look for the first time at the penalties involved when you fail to make a contract. All such penalties are written above the line. They are more severe when you are vulnerable (you have already won a game) than when you are not vulnerable. As we will see in Chapter 16, the opponents may seek to increase the penalties by 'doubling' a contract that they think you cannot make. These are the penalties for undertricks:

Not vulnerable, undoubled	50 per trick
Vulnerable, undoubled	100 per trick
Not vulnerable, doubled	1st trick: 100, 2nd and 3rd tricks: 200, then 300 per trick
Vulnerable, doubled	1st trick: 200, then 300 per trick

Should the declaring side have 'redoubled' the contract (thinking that it could still be made, despite the opponents' double), the penalties would be twice those shown.

Examples

1 You bid Four Spades, non-vulnerable, and make only nine tricks. The opponents score 50.
2 You double the non-vulnerable opponents in Five Clubs and they make only seven tricks, going four down. You score 100 + 200 + 200 + 300, a penalty of 800.
3 You are doubled in Three Diamonds, vulnerable, and make only seven tricks, going two down. The opponents score 200 + 300, a penalty of 500. Had you redoubled the contract, the cost would have been twice as much, 1000 in this case.

We will continue with our imaginary rubber. On the next deal the opponents bid to Four Spades, hoping to make the second game which will give them the rubber. You hold four good trumps and double them. They go two down and, since they are vulnerable, you score a penalty of 500. The scoresheet is now:

We		They
500		
30		
60		100

On the next deal, you bid Four Hearts and make an overtrick. Suppose you had been scoring, yourself. Could you have filled in the score correctly?

This is how the scoresheet should look now:

We	They
30	
500	
30	
60	100
120	

Since you made game, it would also be acceptable to write 150 below the line instead of making two separate entries for the 120 and the 30.

Slam and Rubber Bonuses

Breaking away once more, it is time for us to look at the bonuses you score for bidding and making a slam:

Small slam, not vulnerable	500
Small slam, vulnerable	750
Grand slam, not vulnerable	1000
Grand slam, vulnerable	1500

There is also a bonus for winning a rubber (scoring two games):

When you win 2-0 in games	700
When you win 2-1 in games	500

Back to our imaginary rubber. You and your partner now pick up two enormous hands and bid to 7NT. This is easily made and all that remains is the pleasant task of working out the score. The trick score will be 40 for the first trick and six more tricks at 30, a total of 220. You will also score a vulnerable grand slam bonus of 1500 and a rubber bonus of 500. The final scoresheet will have this, very pleasing, appearance:

We	They
500	
1500	
30	
500	
30	
---	---
60	100
---	---
120	
---	---
220	

Your side has scored 2960, the opponents 100. You have won by 2860. For the purposes of settling the rubber, this is rounded to the nearest 100 (50 being rounded up in the USA, down in Europe). If, in the USA, you were playing for the stakes of 1 cent a point, the opponents would now pay you $29 each. In the UK the word 'point' is (strangely) used to mean a 100 points. If you were playing for 50 pence a point, you would each receive £14.50. Of course the stakes can be as big or as small as you like. Or you can play without stakes.

Scoring honours

A bonus, known as 'honours', is scored above the line when any player holds at least four of the trump honours (A K Q J 10). The bonus is 100 for four honours, 150 for all five honours. In no-trumps a bonus of 150 is awarded when any player holds all four aces.

Bonuses for making a doubled contract

We have not yet mentioned the score you make when the opponents double and you manage to make the contract. The score for all tricks bid and made is doubled (going below the line). You receive also a bonus of 50 non-vulnerable or 100 vulnerable, known as the 'insult'. Each overtrick is worth 100 non-vulnerable, 200 vulnerable.

Examples

1 You are doubled in 3NT, not vulnerable, and make an overtrick. The normal trick value of 100 (40 + 30 + 30) becomes 200. You receive an insult bonus of 50, also 100 for the overtrick.

2 You are doubled in Two Spades, vulnerable, and make the contract exactly. The normal trick value of 60 becomes 120 – enough for game! You receive an insult bonus of 100.

When a rubber cannot be completed

If a rubber cannot be finished for any reason, a side that has made one game scores 300; a side that has a part score in an uncompleted game scores 50.

Quiz

a What entry would be made on the scoresheet if you bid Three Diamonds and made ten tricks?

b What entry would be made on the scoresheet if, on the first deal of a rubber, your opponents bid and made Six Spades?

c What is the purpose of drawing a line across the scoresheet at a certain stage in the play?

d What is the bonus for winning a rubber by two games to one?

e What honours bonus would a side receive if one player held A Q J of trumps and his partner K 10?

Answers

a On the 'We' side of the sheet you would enter 60 below the line and 20 above the line.

b On the 'They' side of the sheet you would enter 180 below the line and 500 (non-vulnerable small slam bonus) above the line.

c To show that one of the sides has scored game.

d 500.

e None. The bonus does not apply unless at least four honours are held by a single player.

12 | STRONG TWO-LEVEL OPENINGS

Sometimes you pick up a hand so powerful that you would be nervous to open with a one-bid. Partner might pass with four or five points and you could then miss game.

These are the types of hand we have in mind:

(1) ♠ A 5
♥ K Q J 10 7 6
♦ A Q J 9 4
♣ –

(2) ♠ K 10 7
♥ A K J 3
♦ A J 2
♣ A Q 6

Suppose you open 1♥ on hand (1) and everyone passes. You will almost certainly have missed a game. Even if your partner holds nothing much of value, it is likely that you can make Four Hearts. You will probably lose only one spade, one heart and one diamond. Even if your partner did respond to an opening of 1♥, you would find it impossible thereafter to express the true value of your hand.

Hand (2), with 22 points, causes the same problem. If you open 1♣ (playing five-card majors) or 1♥ (playing four-card majors), your partner will pass with four points or so and you might then have missed a good 3NT.

What is the solution for these strong hands? You open at the two level! You would in fact open 2♥ on (1), showing a very strong hand with a good heart suit, and 2NT on (2). We will look at the various two-level openings in turn.

The 2NT opening

We have discussed in previous chapters how you bid balanced hands in the ranges 12–14, 15–17, and 18–19. When you hold a balanced hand that is even stronger, you must open at the two level. An opening bid of 2NT shows 20–22 points. It is desirable, but not essential, to have a stopper in each suit.

These are typical 2NT openings:

(1)	♠ K 5	(2)	♠ A Q 5	(3)	♠ K Q 10 7 4
	♥ A Q 10 7		♥ 10 7		♥ A K 2
	♦ K J 5		♦ A K Q J 8		♦ Q 5
	♣ A K J 5		♣ A 10 5		♣ A Q J

Hand (1) is typical, with stoppers in every suit. Hand (2) has no heart stopper but 2NT is still the best description of the hand. Nor is the presence of a five-card major any barrier. Open 2NT on hand (3).

Responder needs around four points to advance to game. He may also want to investigate a major-suit fit. These are the possible responses:

Pass	Fewer than four points.
3♣	Stayman Convention, asking for a four-card major.
3♦/3♥/3♠	Forcing, with at least a five-card suit.
3NT	Enough to raise and no interest in the majors.
4♥/4♠	A six-card suit at least.

Suppose your partner opens 2NT and you hold:

(1)	♠ 6 5 2	(2)	♠ Q 3	(3)	♠ K 10 9 7 4
	♥ 7		♥ K Q 9 7		♥ 2
	♦ K J 9 4 2		♦ 10 8 2		♦ J 9 6
	♣ 10 8 3 2		♣ 9 8 7 4		♣ Q 10 6 3

On (1) you would raise to 3NT. As you will recall, it is nearly always better to try for nine tricks in no-trumps, rather than eleven in a minor-suit game.

On (2) you would bid 3♣ (Stayman) to discover if there was a 4–4 heart fit.

On (3) you have a five-card major and enough strength for game. You show this by responding 3♠, which is forcing. Your partner will usually choose to play in 3NT when he has only two spades, or in 4♠ when he has three or more spades.

Opening Two Diamonds, Two Hearts and Two Spades

An opening bid of Two Clubs has a special meaning, as we will see in a moment. Two of any other suit shows a strong hand with a good holding in the bid suit. These are typical hands:

	(1)	♠ A Q J 10 6 3	(2)	♠ A
		♥ 5		♥ 3
		♦ A K 6		♦ K Q J 10 8 3
		♣ K Q 6		♣ A K 10 7 6

You would open 2♠ on the first hand, 2♦ on the second. These openings are forcing. Unless your partner has a fit or a good suit of his own, he must make the 'negative response' of 2NT. This will allow you to bid again. You would rebid 3♠ on (1), which your partner could pass if he held very little. On (2) you would rebid 3♣, showing your other suit.

A mistake many inexperienced players make is to open with a two bid just because they have 18 or 19 points. Look at this hand:

	(3)	♠ A Q 9 8 5 3
		♥ A 3
		♦ K 9 3
		♣ A Q

Plenty of points, yes, but if your partner has a poor hand with perhaps a singleton or doubleton spade you could lose six or seven tricks. It would be a mistake to open 2♠, forcing the bidding to 3♠ even when your partner has nothing. Open 1♠ instead. If partner cannot respond 1NT or 2♠ it is most unlikely that you will miss a game. The benefit will come when your partner has very little. You can then play at the one level.

This is a typical bidding sequence involving a strong two opening:

West	*East*	*West*	*East*
♠ A J 3	♠ Q 9 4 2	2♥	2NT
♥ A K Q 6 4 3	♥ 9 8 2	3♥	4♥
♦ 4	♦ K 10 7 3		
♣ K Q 3	♣ 10 6		

East gives a negative response and his partner rebids 3♥, non-forcing. Since East has three trumps and a couple of cards that might prove useful, he advances to game.

Any response other than 2NT shows positive values. In particular, a single raise (for example, 2♥ – 3♥) is *not* a limit bid; it is unlimited and will often be a prelude to bidding a slam. As we saw in the sample bidding sequence above, when you have support and just the odd card or two you begin with a negative 2NT.

Opening Two Clubs, unbalanced hand

The two-bids we have seen so far covered hands that were near to game
but needed a little help from the partner before game could be made. What
if you have enough to make game in your own hand? You then open Two
Clubs, the strongest opening bid possible. This is a conventional bid,
saying nothing at all about your clubs. Nearly always you will have 20
points or more when you open Two Clubs. These are typical hands:

(1) ♠ A K J 10 6   (2) ♠ Q
    ♥ A K Q 4 3   ♥ A K Q 9 8 2
    ♦ 4   ♦ K Q 8
    ♣ A 5   ♣ A K 6

With nothing special to show, your partner makes the negative response of
2♦. On (1) you would rebid 2♠, intending to show your hearts on the next
round. The bidding must continue to game, which is the very reason you
opened 2♣. On (2) you would rebid 2♥.

When you have a powerful hand with clubs as the main suit, there is no
'strong two' available. You have to choose between opening a full-blooded
Two Clubs (intending to rebid in clubs) or a simple One Club.

Responding to Two Clubs

If your partner has a good suit to show, or has upwards of nine points, he
may give a 'positive response' (any bid other than 2♦). Suppose your
partner opens 2♣ and you hold one of these hands:

(1)   ♠ 4   (2)   ♠ K 4 3   (3)   ♠ A 9 7
      ♥ A K J 4 3   ♥ Q 9 7 6 3   ♥ 6 2
      ♦ 10 7 6 2   ♦ J 8 2   ♦ K 10 9 4
      ♣ 9 8 4   ♣ Q 3   ♣ Q 8 7 2

On (1) you would respond 2♥, showing your good heart suit. Hand (2) has
the same number of points but the hearts are nothing special; you would
give a negative response of 2♦, waiting to hear what your partner has.
Hand (3) would justify 2NT, a positive response in no-trumps.

Opening Two Clubs, balanced hand

There is one other occasion when you open 2♣. It is when you have a
balanced hand that is too strong for a 2NT opening. This is the scheme:

Open 2♣ and rebid 2NT	23–24 points
Open 2♣ and rebid 3NT	25–27 points
Open 2♣ and rebid 4NT	28–30 points

When the bidding starts 2♣ – 2♦ – 2NT, your partner is allowed to pass when he has nothing at all. This is the only time that the bidding can stop short of game after a 2♣ opening. If instead your partner decides to bid on, he can look for a major-suit fit in the same way as opposite a 2NT opening.

Points to remember

1 An opening of 2NT shows 20–22 points. The responder may bid a Stayman 3♣ to look for a 4–4 fit in a major. Three-level responses in any other suit show a five-card suit at least and are forcing to game.

2 An opening of 2♦, 2♥ or 2♠ shows a hand that requires very little opposite to make game, the sort of hand where you would be nervous of missing game if you opened with a one-bid. These bids are forcing. If your partner has nothing special to show he makes the negative response of 2NT.

3 With an even stronger unbalanced hand, where you want the bidding to proceed to game however weak your partner is, you open 2♣. The negative response to this is 2♦.

4 You open 2♣ also when you hold a balanced hand of 23 points or more. You rebid 2NT with 23–24 points, 3NT with 25–27 points.

Quiz

a What is the main reason for opening 2♥ on a strong hand with hearts, rather than 1♥.

b Is a strong two opening forcing?

c Why is there no strong two available in clubs?

d How would you plan the bidding with a flat hand of 26 points?

e After 2♣ – 2♦ – 2♥, must the bidding continue to game?

Answers

   a Because you fear that a 1♥ opening might be passed out and you would then miss a game.

   b Yes. However weak you are, you are not allowed to pass. You must give the negative response of 2NT.

   c Because 2♣ has a conventional meaning, namely that your hand is very strong indeed - usually worth game in its own right.

   d You would open 2♣, intending to rebid 3NT. This sequence shows a balanced hand of 25–27 points.

   e Yes. The bidding must always continue to game after a 2♣ opener, except for the one sequence 2♣ – 2♦ – 2NT.

13 | OPENING BIDS OF THREE AND FOUR

The time to open with a bid of Three or Four is when you hold a fairly weak hand with a long suit. You hope to buy the contract before the opponents have had the chance to exchange information. Such openings are known as 'pre-emptive bids'. The high-card strength will generally be insufficient for a normal one-level opening. Vulnerability must be taken into account because when you open with a bid of this kind there is always the risk of incurring a penalty.

Opening Three bids

Non-vulnerable, in the first seat, this is a typical hand for an opening bid of Three Spades:

♠ K Q J 9 8 6 4
♥ 5
♦ 10 4 3
♣ J 5

If spades are trumps you expect to make six tricks. If instead the other side chooses trumps, your hand will be almost worthless. Should one of the opponents hold a singleton spade, you might not score any tricks at all in defence.

Suppose you are doubled in Three Spades and your partner does not provide a single trick. Three down doubled, non-vulnerable, costs 500. This is very far from being a disaster because the opponents would surely have made a game at least, had they been allowed to choose trumps. (At rubber bridge they might then enter only 120 or 150 on the scoresheet, but the hidden value of the game makes the true value at least 500.) They might even have made a slam, in which case you have obtained a very good bargain.

It happens more often, when you make this type of opening, that your partner will hold a trick or two and you will be only one or two down, still saving a game the other way. Quite often, too, the opponents will compete. Starting their bidding at such a high level, they may well end up in the wrong spot.

The purpose of opening with a pre-emptive bid is to make life difficult for the other side. Consider such a bid when you have a long and strong suit but relatively few points – you have the type of hand which will make several tricks if you choose trumps, hardly any if the opponents are allowed to choose trumps.

Pre-empts when vulnerable

When you are vulnerable you need to be a bit careful, because three down doubled is 800 – more than a game is worth to the opposition. Still, when the score is Game All it is not a disaster to lose 800 to save a vulnerable game worth around 650.

Position at the table

Another point to bear in mind, when considering a borderline pre-empt, is your position at the table. In first or second position you should not open with a Three bid on any hand that would qualify for an opening bid of One. The main reason is that your partner will expect you to be weak and may pass when your side could easily make a game.

In third position, after two passes, the situation is at its most favourable for pre-emption. Your own hand is weak and neither of the previous players' hands was worth an opening bid. The odds are therefore high that the fourth player has a good hand. You can make life awkward for him by pre-empting. It is also relatively safe to pre-empt with a stronger hand than normal; since your partner has passed, there is not much risk that you will be missing a game.

In third seat, non-vulnerable, you might open Three Hearts on either of these hands:

	(1)		(2)
	♠ 10 4 2		♠ 6 2
	♥ Q 10 9 7 6 3 2		♥ A K J 10 7 3
	♦ 7		♦ K 6 2
	♣ 9 5		♣ 6 5

With hand (1) you know that the fourth player is strong. Even if the two players in front of you both hold a maximum Pass of 11 points, that will still give the fourth player 16 points. Most of the time he will have at least 20 points and will be greatly inconvenienced by your 3♥ opening.

On (2) it is possible that the opponents have a good spade fit. By opening 3♥ you make any such fit difficult to find.

Responding to Three bids

There are two reasons why you might raise an opening of Three Hearts to Four Hearts. We will look first at the situation where you expect to make the game:

West	East	West	East
♠ 5	♠ A J 6 3	3♥	4♥
♥ K Q J 8 7 4 2	♥ A 3		
♦ 10 9 4	♦ K Q 7 3		
♣ 8 4	♣ K 6 2		

East raises to 4♥, which will probably succeed. Look at the strength East needed for this to be a good contract! A pre-emptive opening shows a weak hand and the responder will need several aces and kings to produce ten tricks.

You may also raise to game when you have a good fit for your partner and only moderate strength. The objective in this case is to make life even more difficult for the player on your left.

West	East	West	East
♠ J 5	♠ 10 8 3	3♥	4♥
♥ K 10 8 7 6 4 2	♥ A J 3		
♦ Q 9 4	♦ 7 3		
♣ 8 4	♣ K Q 10 6 3		

Realising that Four Hearts will be relatively cheap, and that the opponents are likely to be able to make game somewhere, East raises the pre-empt. It will now be harder for South to enter the auction. Even if you are doubled and go two down, this will not be a disaster. The opponents could have made Four Spades.

When you hold a strong hand in response, it is important to choose the right game. This will usually be game in the opener's suit, rather than 3NT. Inexperienced players tend to go wrong here:

West	East	South	West	North	East
♠ K J 10 8 6 3 2	♠ 7	–	3♠	Pass	?
♥ 4	♥ A K 7 3				
♦ 10 5	♦ A K 8 2				
♣ 9 7 2	♣ A 10 8 3				

'Since I have no spade support, I'd better bid 3NT,' they say. But 3NT is a hopeless contract. There is no entry to the hand with the spades and the declarer will go at least three down.

Strange as it may seem, East should raise to 4♠. This contract has very good chances. West will probably score five trump tricks, plus five tricks from East's high cards.

Opening Four bids

An opening bid of Four in any suit is also pre-emptive. You need reasonable insurance against a heavy defeat, because the opponents are much more likely to double you in a game bid of Four Hearts or Four Spades than in a part-score contract.

Suppose you hold as dealer:

♠ 6 3
♥ K Q 9 8 7 5 3 2
♦ 10 4
♣ 9

Not vulnerable, you can certainly open Four Hearts. Vulnerable, the resultant penalty might be too heavy. You would open only Three Hearts.

On some hands the risk may be of a different kind - that by opening with a pre-emptive bid you might miss a slam. Suppose you hold:

♠ A K J 9 8 7 4
♥ 5
♦ K J 5 2
♣ 3

If you open Four Spades on a hand this powerful, your partner might pass when he holds enough strength to produce a slam. In the first two seats you should open only One Spade, giving your partner a chance to express his values.

In the third seat, with your partner having passed, you could open Four Spades. The chance of missing a slam is then small and you would be more concerned with making life awkward for the player on your left. The disadvantage with opening only One Spade is that the opponents might then find a fit in hearts or clubs, arriving at a profitable sacrifice.

Points to remember

1 An opening bid at the three level shows a hand with a long suit and insufficient values to open with a one-bid. The intention is to remove bidding space from the opponents.

2 An opening bid at the four level shows a similar hand but with more playing strength, still short of the values for a one-bid.

3 You may raise an opening bid of Three Hearts or Three Spades to game for two quite different reasons. The first is that you expect to make the game (you will need several aces and kings for this, usually around 16 points). The second reason is that you have a good trump fit and wish to make life even more difficult for the opponents.

4 When considering a borderline pre-emptive opening, bear in mind whether you are vulnerable, also your position at the table.

Quiz

a What type of hand is suitable for a pre-emptive opening?

b Why should pre-emptive openings be stronger when you are vulnerable?

c What is the maximum number of points you would hold for a first-seat opening of 3♣?

Answers

a A hand which will make several tricks if you choose trumps, hardly any if the opponents do.

b Because the penalties for going down are higher.

c Around ten points. With any more you would have enough strength to open with a one-bid.

14 | BIDDING SLAMS

To make a small slam you need:

■ the power to make 12 tricks

■ the controls to stop the opponents making two tricks.

By 'controls' we mean aces and kings, singletons and voids. Any of these can prevent the opponents from scoring tricks in a suit.

Suppose you and your partner hold these hands:

West	East
♠ A K Q 8 5	♠ J 10 9 4
♥ A Q 2	♥ K J 9 7 3
♦ Q 10 4	♦ A K
♣ J 4	♣ 10 9

What are the prospects in Six Spades? There is no problem with the 'power'. You have five tricks in spades, five in hearts, and three in diamonds. The problem is that the opponents will be able to score two club tricks before you gain the lead. You are lacking controls, in this case club controls.

This is the other side of the coin:

West	East
♠ A K Q 8	♠ J 10 9 4
♥ A 8 2	♥ K 10 3
♦ K 9 4	♦ A 7 3
♣ A 8 3	♣ K 9 5

You have plenty of controls, including the ace and king of every suit. You are lacking 'power'. In all probability you will lose three tricks, one in each side suit.

Do we have enough power for a slam?

The first step in a slam auction is to determine whether you have enough power to generate 12 tricks. When both hands are balanced, and you are considering a slam in no-trumps, this is largely a matter of points. Roughly speaking, you should bid 6NT when the two hands contain 33 points, 7NT on 37 points.

These two hands contain the values necessary for 6NT:

West	*East*
♠ K Q 4	♠ A J 6 3
♥ Q J 7	♥ K 10 4
♦ K 8 5 2	♦ A Q 3
♣ A 5 3	♣ K J 8

There are 33 points between the hands. By knocking out ♥ A you can set up 11 top tricks. A twelfth will come if the diamonds break 3–3 or if the club finesse succeeds.

A suit slam may be possible on far fewer points, particularly when long suits are present. Look at these two hands:

West	*East*
♠ A 10 9 5	♠ 6 2
♥ A Q J 9 4 2	♥ K 10 5 3
♦ 4	♦ J 3
♣ K 3	♣ A Q J 9 5

Only 25 points between the hands but the power for a slam is there (six heart tricks, five club tricks, and one spade). The controls are there too. West's singleton diamond will stop the defenders from cashing two diamond tricks.

How do you judge if you have enough power for a suit slam? We have seen in previous chapters how you decide if you have enough power to bid a game. As a simple rule, you are entitled to suggest a slam if you hold an ace more than you would need to bid game.

Suppose you are East here:

West	*East*	*West*	*East*
♠ K J 8 3	♠ A Q 9 4 2	1♣	1♠
♥ A 5	♥ K 10 2	3♠	?
♦ Q 4	♦ K 7 3		
♣ A K 9 7 3	♣ Q 5		

You have appreciably more high-card points than you would need to raise to the spade game. There are several other factors which suggest that a slam may be possible:

■ you have good trumps, two honours and a fifth trump
■ you hold an honour in your partner's main suit (♣ Q)
■ you have some sort of control in the other two suits.

It is not just a question of points, as you see. The more you play the game, the more you will appreciate which hands are 'good' and which are relatively 'bad'. To hold good trumps is always a promising sign. A side-suit card such as a queen may be very valuable in your partner's main suit, worth little in some other suit.

Look back to the East–West hands above. Six Spades is a very good contract. East is likely to score five trump tricks, three clubs, two hearts and one diamond. That is 11 tricks and the total can be bumped to 12 by ruffing a heart or a diamond in dummy.

Once East has convinced himself that the power to make 12 tricks is likely to be present, he can use one of the most famous conventions in bridge – Blackwood!

The Blackwood convention

A bid of 4NT asks how many aces your partner holds. These are the responses:

5♣	No aces, or four aces
5♦	One ace
5♥	Two aces
5♠	Three aces

So, on our hand above, the auction would conclude like this:

West	East	West	East
♠ K J 8 3	♠ A Q 9 4 2	1♣	1♠
♥ A 5	♥ K 10 2	3♠	4NT
♦ Q 4	♦ K 7 3	5♥	6♠
♣ A K 9 7 3	♣ Q 5	End	

East hears that his partner holds two aces. Since the opponents do not have two aces to cash, he bids the slam.

Suppose that West held a different hand:

West	East	West	East
♠ K J 8 3	♠ A Q 9 4 2	1♣	1♠
♥ Q 5	♥ K 10 2	3♠	4NT
♦ J	♦ K 7 3	5♦	5♠
♣ A K J 9 7 3	♣ Q 5	End	

This time West shows only one ace. Knowing that the opponents hold two aces, East bids only Five Spades.

After asking for aces with 4NT, you can bid 5NT to ask how many kings your partner holds.

West	East	West	East
♠ Q 8 3	♠ A K J 7 4 2	1♣	2♠
♥ Q J 2	♥ A K 3	3♠	4NT
♦ A J	♦ K 7	5♦	5NT
♣ K 10 9 7 3	♣ A 4	6♦	7♠
		End	

Once East knows that all the aces are present he decides he will bid the grand slam if West holds ♣ K. If West had not held any kings he would have responded 6♣ instead and East would have signed off in 6♠.

Most players use Blackwood too often. The time to use it is when:

■ you know already that the power for a slam is present

■ you hold a control of some sort in each suit.

How do you find out whether you have two top losers in one of the side suits? It is only possible if you use a type of bid which we have not yet seen: a control-showing cue bid.

Control-showing cue bids

When a trump suit has been agreed, and the bidding has reached the four level, a bid in a new suit is a 'control-showing cue bid'. It will usually show the ace or king of the suit bid.

If you think this is rather more complicated than the type of bidding we have seen so far, you are right! Bidding slams accurately is difficult, even for experts. Although you may not wish to use cue bids yourself for a while, you may find it interesting to see how they work. This is a typical auction:

West	East	West	East
♠ K J 10 3	♠ A Q 9 4 2	1♥	1♠
♥ A K J 9 7 3	♥ 8 6	3♠	4♣
♦ J 2	♦ Q 3	4♥	4♠
♣ Q	♣ A K 10 5	End	

Look at the bidding problem East faces when his partner raises to 3♠. Blackwood would be no use. It would tell him that there were not two aces missing but it would not let him know if the diamond suit was controlled.

Instead East bids 4♣. Since spades have been agreed as trumps and the bidding is at the four level, this is a 'control-showing cue bid'. It shows the ace or king of clubs and asks West to bid his cheapest control. If West held the ace or king of diamonds (or a singleton or void in the suit) he would cue-bid 4♦. When instead he cue-bids 4♥, he shows the ace or king of hearts and denies any diamond control. Knowing that the two top diamonds are in the hands of the defenders, East stops in game.

This is how the auction would have gone if West did hold a diamond control:

West	East	West	East
♠ K J 10 3	♠ A Q 9 4 2	1♥	1♠
♥ A K J 9 4	♥ 8 6	3♠	4♣
♦ K 2	♦ Q 3	4♦	4NT
♣ Q 4	♣ A K 10 5	5♦	6♠
		End	

When East hears that the diamond suit is controlled, he bids Blackwood. (True, he holds no heart control, but since this is his partner's main suit it is a near certainty that West will hold a heart control.) West's 5♦ response confirms that there are not two aces missing and he bids the slam.

The mathematics of slam bidding

If you bid a non-vulnerable small slam and make it, you receive a bonus of 500. What if you bid the slam and go one down? Some players shrug their shoulders. 'It only cost 50,' they say.

The actual cost is much more than that, because they have lost the value of the non-vulnerable game they would otherwise have made (worth around 450.) Since you will gain 500 if you make the slam, lose 500 if it goes down, you should bid a slam only if you think it will be better than an 'Evens' chance.

The cost of bidding a failing grand slam is even greater. You lose the value of the game and that of the small slam you could have made. You should bid a grand only if you think it is about a '2–1 on' favourite to make.

Slam bidding is a difficult art and the best advice when you are starting the game is to bid slams only when it seems obvious to do so.

Points to remember

1   To make a slam you need the power to make 12 tricks, also the controls to stop the opponents making two tricks first.

2   To make 6NT with two balanced hands you need a total of around 33 points. To make 7NT you need about 37 points.

3   It is more difficult to judge if you have the playing strength to make a suit slam. Investigate a slam when you have around an ace more than you would need to bid game.

4   When you decide that the power is present for a slam, and the odds are good that every suit is under some sort of control, you may bid a Blackwood 4NT. Your partner's response will tell you whether two aces are missing.

5   Once a trump suit is agreed, any bid in a new suit at the four level or higher is a control-showing cue bid. It shows that you hold the ace or king of the suit (sometimes a singleton or void). This will allow you to discover if one of the side suits is unguarded.

6   When you first play bridge, bid only those slams which are obvious. Remember that if you go down in a slam, the loss is not only the 50 or 100 visible on the scoresheet. You have lost also the value of the game that you would otherwise have made.

Quiz

a   When should you think of bidding a slam?

b   Does the Blackwood convention help you to know if you have the power necessary to bid a slam?

c   What is a 'control-showing cue bid'?

d In this auction, is West's second bid a cue bid?

West	East
1♠	2♠
3♦	

e What is the meaning of East's 4♣ bid in this auction:

West	East
1♦	2♠
3♠	4♣

Answers

a When you have appreciably more strength than you need to bid game.

b No. It tells you only if you have two aces missing.

c After a trump suit has been agreed, a bid in a new suit at the four level (or higher) shows a control in that suit. This will usually be an ace or king, sometimes a singleton or void.

d No, because it is only at the three level. West's 3♦ bid here is a game try in spades (see Chapter 7).

e Spades have been agreed as trumps and East's 4♣ is at the four level or higher. It is therefore a control-showing cue bid.

15 | OVERCALLS

Until now we have considered only the auctions where your side has made the opening bid. Often, however, you will want to contest the auction after the opponents have opened.

Suppose the player on your right opens 1♣ and you overcall 1♠. ('Overcall' means to make a bid once the opponents have opened.) What is the purpose of the 1♠ bid? You are not so much thinking about bidding a game contract, although it may turn out that way; you are principally concerned with making it difficult for the opponents to reach, and achieve, their best contract. With this in mind, we will look first at simple overcalls at the one level.

Simple overcalls at the one level

When overcalling, high cards are not so important as playing strength. Suppose the player on your right opens 1♣ and in second position you hold:

♠ K Q 9 7 6 3
♥ 5 4
♦ 8 3
♣ Q 7 6

You would not open the bidding on such a hand, of course, but there are advantages in overcalling 1♠. It's not particularly likely that your side will bid to a high contract in spades, but you may make life uncomfortable for the opponents. They may be reluctant to bid 3NT with only a modest spade guard. Also, your bid will deprive them of bidding space. The player on your left will not now be able to respond 1♦ or 1♥. It will be more difficult for the opponents to find any diamond or heart fit they may have. A further advantage is that you have suggested a good suit to lead, should your partner have to make the opening lead.

You may also have taken the first steps to a successful 'sacrifice'. A sacrifice is a contract that you expect to go down. You bid it because the penalty you expect to suffer will be less than the value of the contract the opponents would otherwise have made. Here is an example of a 'sacrifice bid':

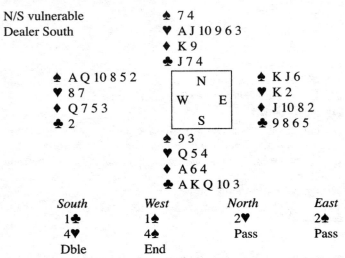

N/S vulnerable
Dealer South

```
                        ♠ 7 4
                        ♥ A J 10 9 6 3
                        ♦ K 9
                        ♣ J 7 4
    ♠ A Q 10 8 5 2         N              ♠ K J 6
    ♥ 8 7               W     E           ♥ K 2
    ♦ Q 7 5 3              S              ♦ J 10 8 2
    ♣ 2                                   ♣ 9 8 6 5
                        ♠ 9 3
                        ♥ Q 5 4
                        ♦ A 6 4
                        ♣ A K Q 10 3
```

South	*West*	*North*	*East*
1♣	1♠	2♥	2♠
4♥	4♠	Pass	Pass
Dble	End		

North would have made Four Hearts, losing only two spade tricks and the king of trumps. Meanwhile, how will West fare in Four Spades doubled? If the defenders are not on their toes, he will lose one heart, two diamonds and a club. Just one down, for a penalty of 100. Even if they take a diamond ruff, putting the contract two down, the penalty will be only 300 – much less than the value of an enemy heart game (around 620).

Another factor in favour of making a sacrifice is that sometimes the opponents will bid one level higher in their own suit. Suppose North or South on this deal had bid 5♥ over 4♠, hoping to make 11 tricks. They would then have gone one down.

How strong do you need to be to make a one-level overcall? The point-count is not important, but the normal range will be about 7–15 when non-vulnerable. Vulnerable, you are risking a heavier penalty, and the minimum would be more like nine points. Even so, it would not be wrong to overcall on a strong suit such as A Q J 9 7 3, with no outside strength.

Sometimes you will hold quite a few points but your long suit will be
weak. Suppose you hear a 1♦ opening on your right and you hold:

           ♠ A Q 4
           ♥ J 9 7 6 2
           ♦ 5 4
           ♣ K 8 3

Ten points, yes, but a poor heart suit. You would not welcome the lead of
such a suit; nor would you want your partner to sacrifice in hearts (your
high cards in the black suits would have been good in defence against their
contract). Since a 1♥ overcall would not rob the opponents of any bidding
space either, it is best to pass.

So, overcall when you have a good suit. The point-count is relatively
unimportant.

Simple overcalls at the two level

At the two level you are much more likely to run into a penalty double; a
fairly strong suit is essential. Suppose at Game All the bidding is opened
on your right with 1♥. Would you overcall on either of these hands?

(1) ♠ A Q 5   (2) ♠ A 5
     ♥ Q 10 8   ♥ 7 4 2
     ♦ A 10 7 5 3   ♦ J 3
     ♣ 6 4   ♣ K Q 10 9 6 2

Hand (1) contains 12 points but the diamond suit is poor. If you overcall
2♦ and the next player doubles, you might go for a very big penalty.
Meanwhile, your top cards may well be enough to prevent the opponents
from scoring game; the penalty would not even be in a good cause. The
wisest course is to pass.

Hand (2) has only 10 points, but a much better suit. It represents a
minimum vulnerable overcall at the two level.

Intermediate jump overcalls

When you hold a six-card suit and enough strength to have opened the
bidding, you can pass this message with an intermediate jump overcall.
Suppose you hear 1♦ on your right and you hold one of these hands:

(1) ♠ A Q J 9 8 3   (2) ♠ 7 6
     ♥ 5   ♥ 5 3
     ♦ A 10 8 3   ♦ A J 4
     ♣ Q 4   ♣ A K J 10 8 4

You would overcall 2♠ on the first hand, 3♣ on the second.

In both cases you have bid one level higher than was necessary. These jump overcalls are not forcing. Your partner will know that you have a good suit of at least six cards and around 12–16 points. Any further advance may be judged accordingly.

The 1NT overcall

Whatever style of opening 1NT you play, weak or strong, you need a strong hand when you overcall 1NT. The player to your left has heard his partner open the bidding and will be quick to double 1NT for penalties when he has some values. You should hold around 15–18 points and a stopper in the opponent's suit.

Over an opening of 1♠ you would overcall 1NT on either of these hands:

(1) ♠ A 9 5   (2) ♠ K J 5
     ♥ A Q 10   ♥ 7 4
     ♦ K J 7 3   ♦ A Q J 8 2
     ♣ Q 8 4   ♣ A J 6

Don't worry about the low doubleton heart on hand (2). 1NT describes your hand much better than an overcall of 2♦. Always choose the bid which will paint the clearest picture of your hand.

Responding to suit overcalls

When your partner overcalls, he is telling you that he has a good suit of his own. He is not asking you to bid one of your suits. Whenever you have a smattering of support you should consider a raise of the overcall. Suppose the bidding starts 1♦ – 1♠ – Pass and you are sitting in the fourth seat with one of these hands:

(1) ♠ Q 9 5   (2) ♠ K 10 9 5   (3) ♠ A J 7 3
     ♥ J 3   ♥ 4   ♥ A Q 7 5
     ♦ K 10 6 4 2   ♦ J 8 3   ♦ 7 5
     ♣ 10 7 2   ♣ K J 8 5 2   ♣ Q 10 3

Raise to 2♠ on hand (1). It is not that you think your side can make 4♠; you want to take bidding space away from the opener, who is likely to hold a good hand. On hand (2), with four-card spade support and a side-suit singleton, you can afford to be even more awkward. Raise to 3♠.

As you see, raises of your partner's suit are obstructive. You are making it harder for the opponents to discover where there own fit lies. When instead you see a genuine chance of game, as on hand (3), you show this by cue-bidding the opener's suit. Here you would bid 2♦, showing that you have a good fit for your partner's spades and you think a game may be possible. Bidding a suit that the opponents have called is a useful mechanism in many situations. It always shows a strong hand.

Without a fit for your partner's suit you should bid only if your hand is quite strong. Remember that the overcaller may not have many points. Suppose again that the bidding has started 1♦ – 1♠ – Pass and you hold one of these hands:

(1)	♠ 4	(2)	♠ 8 2	(3)	♠ 7 3
	♥ Q 4		♥ A 7 2		♥ K J 8 2
	♦ J 8 4 3		♦ 9 8 3		♦ A Q 8 2
	♣ A J 9 8 7 2		♣ A K J 8 2		♣ J 10 5

Pass on hand (1), rather than bidding 2♣. Your partner will hold a five-card spade suit at least and there is no need to fear a contract of 1♠. In any case, you should not bid a new suit merely to rescue your partner. The meaning of a response in a new suit is that you are quite strong and see a possibility of game.

Hand (2) is worth a response of 2♣. If your partner has overcalled on around 13 points, rather than a minimum eight or so, he will realise that there should be a game somewhere. You can respond in no-trumps on (3), but make it only 1NT not 2NT. Remember that your partner has not shown the values for an opening bid, only for an overcall at the one level.

Points to remember

1 An overcall shows a good suit; points are less important. You overcall to suggest a good lead, to consume the enemy's bidding space, and to allow your side to contest the auction. Sometimes you may find a profitable sacrifice against a game by the opponents.

2 An intermediate jump overcall shows a suit of at least six cards and around 12–16 points.

3 A 1NT overcall shows 15–18 points, a fairly balanced hand, and a stopper in the suit bid by the opponents.

4 When responding to a suit overcall, raise whenever possible. Even a jump raise is obstructive in purpose. When instead you see a genuine chance for game by your side, cue-bid the opponents' suit.

5 When your partner has made an overcall, do not bid a new suit as a rescue. A change of suit is constructive and suggests that game may be possible.

Quiz

a With neither side vulnerable, the player on your right has opened 1♦. What call would you make on each of these hands?

(1)	♠ A K 9 6 3	(2)	♠ K Q 10 8 7 2	(3)	♠ Q 5 3
	♥ 9 2		♥ A 7 2		♥ J 9 5 4 2
	♦ 8 5 3		♦ 9 3		♦ K Q 2
	♣ J 9 2		♣ A 5		♣ K 8

b With the opponents vulnerable, the bidding starts 1♣ – 1♥ – Pass. What response, if any, would you make to your partner's overcall on each of these hands?

(1)	♠ 9 8 2	(2)	♠ A K 4 3	(3)	♠ K Q 5 3
	♥ A 7 3		♥ J 9 7 2		♥ 3
	♦ Q J 8 7 2		♦ 9 3		♦ 6 2
	♣ 9 6		♣ A 5 2		♣ J 9 7 6 5 2

c What does the term 'sacrifice' mean?

Answers

a On (1) you should overcall 1♠. You have a good suit and you will stop the next player from responding 1♥. On (2) you are worth an intermediate jump overcall of 2♠. On hand (3) you would pass. The heart suit is poor and an overcall would not deprive the opponents of any bidding space.

b On (1) you should raise to 2♥, aiming to make life more difficult for the opener. On (2) you have a chance of game your way; show your strong support with a cue bid of 2♣. With hand (3) you should pass. To bid a new suit would be a constructive move.

c A sacrifice is a contract that is bid in the expectation of going down. You hope that the penalty will be less than the value of the contract the opponents would otherwise have made.

16 | PENALTY AND TAKE-OUT DOUBLES

At any time when the most recent bid was made by an opponent, you are allowed to 'double'. When the game of bridge was first conceived, this call had only one meaning – that you thought the opponents could not make their contract and you wanted to increase the stakes so that the penalty would be larger.

This type of double is known as a 'penalty double'. Suppose you are West and hear this auction:

West	South	West	North	East
♠ Q J 9 2	1♠	Pass	2♠	Pass
♥ A 5 4	4♠	?		
♦ A K 7 6				
♣ 10 3				

You might think to yourself: I may well score two trump tricks and I have three possible winners outside; I don't think they are going to make this. In that case you say 'Double'. If the contract does goes down, the penalties will be increased, as we saw in the chapter on scoring. Should the declarer make the contract, perhaps because he is very short in diamonds and your ♦ A K are unproductive, the opponents' score will be increased instead.

It was soon realised that in many situations, particularly at a low level, a better use could be found for a double. Instead of attempting to penalise the opponents, the call would mean 'I have a good hand but no good suit to bid; what is your best suit, partner?' This type of double is known as a 'take-out double'.

Suppose the player on your right opens 1♥ and you hold this hand:

West	South	West	North	East
♠ A 10 7 2	1♥	?		
♥ 5				
♦ K J 8 4				
♣ K Q 9 2				

You have enough to enter the auction, but no suit good enough for an overcall. You double, for take-out. Your partner can respond to the double with a bid in any of the other three suits. Perhaps your partner has a hand such as:

South	West	North	East		East
1♥	Dble	Pass	2♦		♠ K 6 3
					♥ J 8 4
					♦ Q 10 6 3
					♣ J 5 4

He responds in his longest suit, here diamonds, and you will have found your 4–4 diamond fit.

These take-out doubles were found to be so useful that nowadays most doubles are for take-out rather than for penalties. It is important to know which are which, of course.

Distinguishing between take-out and penalty doubles

In general, a take-out double asks your partner to indicate where his strength lies. Consequently, any double made *after* your partner has opened the bidding is a penalty double.

South	West	North	East
1♥	2♦	Dble	

North's double is for penalties because his partner has opened the bidding. Many of the biggest penalties come from low-level contracts. Here West's overcall is a step in the dark, made in the hope that his partner will hold some values. When North holds good diamonds and East is weak, the penalty will often be as high as 800.

A double would also be for penalties after two bids by your side:

South	West	North	East
1♥	Pass	1♠	2♦
Dble			

South has something good in diamonds and expects East's contract to fail.

A double of an opening 1NT bid is for penalties, too:

South	West	North	East
1NT	Dble		

West has a strong hand and expects 1NT to fail. East should now pass, unless he is very weak and has a suit of at least five cards.

The only other situation in which a double is for penalties is the one we saw at the start: the opponents bid to some high contract and you double because you have an unexpectedly good trump holding.

At all other times a double is for take-out. There are countless situations where such a call is useful.

East	South	West	North	East
♠ A 7	1♥	Pass	1♠	Dble
♥ 5 3 2				
♦ A K 8 4				
♣ K J 6 2				

East's double is for take-out and asks his partner to choose one of the unbid suits, here clubs or diamonds.

East	South	West	North	East
♠ A Q 7 2	1♥	Pass	3♥	Dble
♥ 5				
♦ K Q J 7 3				
♣ A J 6				

Even though the bidding has risen to 3♥, East's hand is strong enough to justify a take-out double.

A take-out double is useful also when an opponent has made a pre-emptive opening:

East	South	West	North	East
♠ A K J 2	3♦	Dble		
♥ A 10 9 4 3				
♦ 7 3				
♣ A 6				

There are two problems with the alternative call of 3♥. Your hearts are not good enough for an overcall at the three level. Also, you may miss a good fit in spades. A take-out double is easily the best option. Your partner will expect you to be more interested in the major suits than in clubs.

The requirements for a take-out double

The most common situation for a take-out double is when the bidding has been opened on your right. If you have a fair hand, shortage in the bid suit, and can cope with any response that your partner may make, you may launch a counter-attack by doubling.

Suppose your right-hand opponent opens 1♦ and you hold:

(1)	♠ K 9 7 5 2	(2)	♠ K 4
	♥ A J 8 4		♥ A 9 6 3
	♦ 5		♦ K 8 4
	♣ Q 10 3		♣ K 6 3 2

On (1) your point count is low for a double, which normally shows the values for an opening bid. Since you have good support for both majors, you should go ahead with a double. Hand (2) has more points but the support for spades is poor. On this hand it would be wiser to pass.

Take-out double on a strong hand

When your partner hears a take-out double he will assume initially that you have a hand of the most common type – opening bid values and short in the opener's suit. However, there is another type of hand on which you start with a double: a hand that does not contain support for all the unbid suits but which is too strong for an overcall.

Suppose the player on your right opens 1♣ and you hold one of these fine hands:

(1)	♠ A K Q 7 5 2	(2)	♠ A 4
	♥ 8 4		♥ A K J 6 3
	♦ A K 4		♦ K Q J 4
	♣ K 3		♣ 9 4

Hand (1) is much too strong for an overcall of 1♠ or even 2♠. You would double first and then bid spades over your partner's response. Such an action denotes a good hand. The same is true on hand (2). Although you have no support for spades you would start with a double. If your partner responds 1♠ you will continue with 2♥, showing a strong hand.

Responding to a take-out double

When you respond to a take-out double you have two duties: to show your best suit (or to choose no-trumps), also to give some indication of your strength. Even though the opponents have opened the bidding, your side may still be able to make game. If you have a strong hand facing a take-out double, you must let your partner know the good news.

This is the general scheme of responses:

0–7 points	Bid your best suit at minimum level
8–10 points	Jump in your best suit (non-forcing)
11+ points	Bid game, or cue-bid opponent's suit

If you have good stoppers in the opponent's suit you may respond in no-trumps:

5–9 points with stopper(s)	Respond 1NT
10–12 points with stopper(s)	Respond 2NT
13+ points with stopper(s)	Respond 3NT

Suppose the bidding has started 1♦ – Dble – Pass and you have to find a response on one of these weakish hands:

(1)	♠ J 9 7 5	(2)	♠ 9 8 7	(3)	♠ 8 3
	♥ 10 3		♥ J 6 3		♥ 9 5 2
	♦ 9 8 3		♦ J 9 8 4		♦ A J 10 4
	♣ A 8 7 3		♣ 10 8 4		♣ Q 6 3 2

On (1) you would respond 1♠, showing the major suit rather than the minor. Hand (2) is awkward. Do not make the mistake of passing! Your partner is likely to be short in diamonds and the contract of One Diamond doubled will easily be made, usually with overtricks. Your partner has asked you to choose a suit and it is not your fault that you do not have a four-card suit to bid. Respond 1♥. On hand (3) you respond 1NT, showing 5–9 points and at least one stopper in the enemy suit.

When your hand is in the middle range of 8–10 points, there may be a game on. Remember that if you make a response at the minimum level, you may have almost nothing, as on hand (2) above. It would not be sensible to make the same response on a healthy nine-point hand. Suppose again that the bidding has started 1♦ – Dble – Pass and you now hold a hand in the middle range:

(4)   ♠ A Q 9 3   (5)   ♠ 10 3   (6)   ♠ J 3
     ♥ 9 5   ♥ 9 8   ♥ 9 5 2
     ♦ A 8 3   ♦ 10 7 2   ♦ K Q 9 7
     ♣ 10 9 7 3   ♣ A K J 8 6 2   ♣ A J 3 2

Respond 2♠ on hand (4). There is no need to fear advancing to the two level; your partner has the values for an opening bid, remember. On (5) respond 3♣. If your partner has a diamond stopper, he may be able to make 3NT. On (6) you would respond 2NT, showing around 10–12 points and a good stopper in diamonds.

When you hold close to an opening bid yourself, or a long major suit, you are entitled to head for game. After a start of 1♦ – Dble – Pass, you might hold:

(7)   ♠ 10 3   (8)   ♠ Q 3   (9)   ♠ K Q 8 3
     ♥ A Q J 9 4 2   ♥ 9 8   ♥ A K 9 2
     ♦ 9 8 4   ♦ A J 8 3   ♦ J 8 4
     ♣ K 4   ♣ A K 9 7 2   ♣ 9 6

On (7) you would respond 4♥. The fifth and sixth cards in hearts are worth at least three points and you will be able to make ten tricks opposite most minimum take-out doubles. On hand (8) you would respond 3NT. No need to worry that you are weak in hearts; your partner's double of 1♦ suggests that he will have good values in both the majors.

Hand (9) presents a small problem. You want to be in game, yes, but which game? If you simply guess, jumping to either 4♥ or 4♠, you may guess wrongly and land in a suit where your partner does not have four-card support. There is a clever solution available to the problem. You respond to the double by making a strength-showing cue bid in the opponent's suit. Here you would respond 2♦. The whole auction might then go like this:

West	East	South	West	North	East
♠ A J 10 2	♠ K Q 8 3	1♦	Dble	Pass	2♦
♥ Q 7 4	♥ A K 9 2	Pass	2♠	Pass	4♠
♦ 7 3	♦ J 8 4	End			
♣ A K 8 3	♣ 9 6				

You find the 4–4 spade fit and arrive in the best game. Four Hearts would not be a safe contract, with only seven trumps between the hands.

Leaving in a take-out double

You are allowed to pass when your partner has made a take-out double.
You should do so only when you are very strong in the opponent's suit and
are therefore happy to make their suit trumps.

East	South	West	North	East
♠ 7 3	1♦	Dble	Pass	Pass
♥ 5 2				
♦ Q J 10 8 6 2				
♣ J 8 5				

East passes the take-out double, hoping that the contract will go down.

Redouble by the third player

Sometimes an opponent will choose the wrong moment for a take-out
double. You will be sitting over him with a strong hand:

North	South	West	North	East
♠ A Q 7 6	1♦	Dble	Rdble	
♥ K 10 3				
♦ J 4				
♣ J 10 7 4				

Your partner opens 1♦ and West doubles. When you have upwards of ten
points and no particularly long suit, you can redouble. The meaning is:
I have a good hand; we may be able to double the opponents.

This is the most common use of the redouble call. The other situation
occurs at the game level. Suppose an opponent has doubled you in Four
Spades and you think the contract can be made nevertheless. You are
entitled to redouble, increasing the stakes yet further.

Points to remember

1 You may make a penalty double when your side has opened
   the bidding and an opponent has overcalled.

2 A double of 1NT is for penalties, as is a double of any game
   contract bid by the opponents. To double a suit game such as
   Four Spades, you should have a good trump holding.

3 All other doubles are for take-out. The meaning of a take-out
   double is 'I have a good hand but no good bid to make.
   Please choose a trump suit.'

4 When responding to a take-out double, you must choose a trump suit, or no-trumps; you must also give some indication of your strength. With 0–7 points and no very long suit, respond at the minimum level. With 8–10 points, jump one level. With 11 or more points, or a long major suit, head for game.

5 When you have enough for game but do not know which game will be best, respond to the take-out double with a cue bid in the opponent's suit.

Quiz

a The player on your right opens 1♦. What would you bid on each of these hands:

(1)   ♠ J 10 3   (2) ♠ K Q 4   (3) ♠ A K J 8 3
      ♥ K Q J 9 4   ♥ A 9 8   ♥ A Q 9 2
      ♦ A 8 4   ♦ A J 8 3   ♦ 8
      ♣ K 4   ♣ K 8 3   ♣ A 7 3

b The bidding starts 1♦ – Dble – Pass. How would you respond to your partner's take-out double on each of these hands:

(4)   ♠ A 10 8 7 3   (5) ♠ 9 2   (6) ♠ J 5
      ♥ 8 2   ♥ 8 5   ♥ A K 9 2
      ♦ A J 4   ♦ J 9 7 6 2   ♦ Q 7
      ♣ 9 7 5   ♣ J 8 4 3   ♣ K J 8 7 2

Answers

a On (1) you would overcall 1♥. Remember that a take-out double means 'I have a good hand but no good bid to make.' Here you do have a good bid to make. On (2) you would overcall 1NT. Only hand (3) is a sound take-out double.

b On (4) you would respond 2♠. This is non-forcing but shows about 8–10 points. On (5) you would respond 2♣. Do not make the mistake of passing; the declarer is almost certain to make One Diamond doubled. On (6) you have enough for game but you don't yet know which game. Respond with a strength-showing cue bid of 2♦.

17 | BASIC MOVES IN DEFENCE

It is much easier to do well as declarer than as a defender. That's because the declarer can see all the cards at his disposal; the defenders do not know which cards their partner holds. There are several general guidelines to assist the defenders and in this chapter we will look at them in turn.

Second hand plays low

When you are second to play to a trick, it is usually right to play low. To show why this is a good idea, let's see a few situations where playing high would cost a trick.

North
♦ Q 8 3

West   *East*
♦ A 10 6 4   ♦ J 9 5

South
♦ K 7 2

Declarer leads ♦2 from his hand and you hold the West cards. If you mistakenly play high, rising with the ace, declarer will subsequently score two tricks – one with the queen, another with the king. If you follow the rule of 'second hand low', dummy's queen will win the first trick. Declarer will not make a second trick with the king, however; your ace lies over it.

This position is similar:

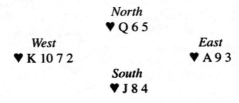

North
♥ Q 6 5

West   *East*
♥ K 10 7 2   ♥ A 9 3

South
♥ J 8 4

Declarer leads ♥4 from his hand. If you rise with the king, declarer will eventually make a trick with one of his honours. If you play low, your partner will win dummy's queen with the ace and declarer will make no tricks at all from the suit.

'But what if declarer held ♥A?' you may be thinking. 'Then I'd do better to play the king.'

That's why defending can be difficult; you don't actually know what cards the declarer has. In this case, though, the declarer would score two heart tricks anyway if he held ♥A. So, playing low would gain in one situation, not cost in the other.

Another reason for playing low in second seat is that you may give the declarer a guess in the suit. Suppose the declarer has this side suit in a trump contract:

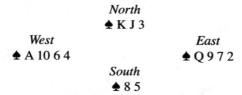

North
♠ K J 3

West
♠ A 10 6 4

East
♠ Q 9 7 2

South
♠ 8 5

Declarer leads ♠5 from his hand. If you rise with the ace, your side will make only one spade trick. Dummy's king will win the second round and declarer will be able to ruff any further spade leads.

If instead you play low, the declarer will have to guess which spade to play from dummy. If he guesses wrong, playing the jack, your side will score two spade tricks.

Of course, it would give the situation away if you thought about playing the ace and then played low. To give the declarer a guess you must play low straight away, following the guideline of 'second hand low'.

Many inexperienced players go wrong when there is a tenace sitting over one of their honours. (A tenace consists of two non-touching honour cards, such as A Q or K J.) Look at this position:

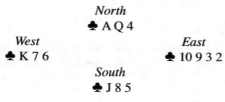

North
♣ A Q 4

West
♣ K 7 6

East
♣ 10 9 3 2

South
♣ J 8 5

When the declarer leads ♣5, they very unwisely play the king. As a result, the declarer makes three club tricks. The excuses such players make are, 'My king was dead anyway, under the ace-queen,' or 'I wanted to force a high card from dummy.'

There is no sense in either of these remarks. If West follows the maxim 'second hand low', the declarer will finesse the queen successfully and make a second trick with the ace. That will be all, however. West's king will win the third round.

Third hand plays high

The situation is different when you are third to play to a trick. Now you must play high, attempting to win the trick. No matter if the fourth player can beat your card. You will at least have stopped him from winning the trick cheaply.

This is a typical situation:

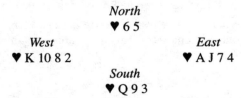

```
                        North
                        ♥ 8 5
        West                            East
        ♥ Q 10 6 3                      ♥ K 9 7 4
                        South
                        ♥ A J 2
```

You are sitting East and your partner leads ♥3 to the trick. If you fail to play the king (third hand high), the declarer will score an undeserved trick with the jack. Play the king, following the rule 'third hand high', and the declarer will make only one trick from the suit.

There are many similar positions:

```
                        North
                        ♥ 6 5
        West                            East
        ♥ K 10 8 2                      ♥ A J 7 4
                        South
                        ♥ Q 9 3
```

West leads ♥2. If East plays the jack, the declarer will make a trick with the queen. East should of course play the ace – third hand high. Declarer will not then make a heart trick.

The rule does not necessarily apply when dummy holds an honour and you hold a higher honour.

North
♠ K 8 5

West   *East*
♠ J 9 6 2   ♠ A Q 4

South
♠ 10 7 3

Your partner leads ♠2 and the declarer plays low from dummy. You win the trick with the queen, not the ace (which would make dummy's king a winner).

Less obvious is this situation:

North
♠ K 7 4

West   *East*
♠ Q 9 8 2   ♠ A J 3

South
♠ 10 6 5

Your partner leads ♠2, dummy playing the 4, and you must play the jack, not the ace. Perhaps you would be worried that the declarer held the queen and the actual position was like this:

North
♠ K 7 4

West   *East*
♠ 10 8 6 2   ♠ A J 3

South
♠ Q 9 5

It would still be right to play the jack! Declarer would win with the queen but this would be his last trick from the suit. If instead you played the ace on the first round, the declarer would score two tricks.

The way to look at it is this. Your ♠A has a purpose in life – to capture a high card, such as a king or a queen. If you play it while ♠K is still in dummy, the declarer will eventually make a trick with the king. If you keep hold of the ace, dummy's king will not score a trick.

Touching honours in third seat

When you are third to play to a trick and have two high cards in sequence (such as A K, K Q or J 10), you should play the *lower* of the two cards. This will help your partner to read how the cards lie. Suppose West leads ♦3 here:

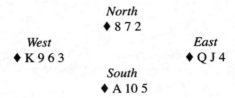

North
♦ 8 7 2

West
♦ K 9 6 3

East
♦ Q J 4

South
♦ A 10 5

You should play the jack, the lower of touching cards. When the declarer wins with the ace, your partner will know that you hold the queen. (Otherwise the declarer would have won with the queen, making two tricks from the suit.) If you failed to follow this scheme, and played the queen at trick 1, your partner would have no idea where the jack was.

Be wary of starting a new suit

In most cases it is a big disadvantage to have to make the first lead in a suit. There are countless situations where it will cost a trick.

North
♠ Q 7 4

West
♠ K 10 6 2

East
♠ A 9 5

South
♠ J 8 3

Suppose West makes the first play in this suit. East will have to play the ace, to stop South's jack from winning, and now the declarer will make a trick on the third round. The position is the same if East leads to the trick. If instead the declarer has to make the first lead, he cannot score a trick.

It's the same here:

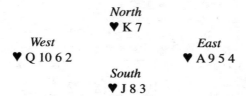

North
♥ K 7

West
♥ Q 10 6 2

East
♥ A 9 5 4

South
♥ J 8 3

If West or East make the first lead, the declarer will score a trick. If instead the declarer plays first, leading to the king, East will take the king with the ace and West will win the second and third rounds.

This is yet another situation:

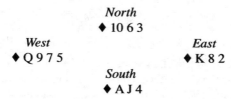

North
♦ 10 6 3

West
♦ Q 9 7 5

East
♦ K 8 2

South
♦ A J 4

If West leads the suit, East will have to play the king to stop South's jack from winning. Declarer will capture with the ace and will make a second trick from his jack and 10. It's the same if East makes the first play, leading the 2. Declarer will play low from his hand and West will win with the queen. Now the declarer will score two tricks with the A J sitting over East's king. If the declarer has to play the suit himself, there is no way to make two tricks. Try it!

Why have we made this point at such length? Because the most common mistake that inexperienced players make in defence is to 'try a new suit' in the middle of the hand. Much of the time they will be giving away a trick by doing so.

Let's see how this applies to a full deal.

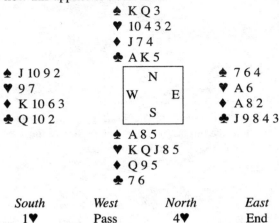

```
                    ♠ K Q 3
                    ♥ 10 4 3 2
                    ♦ J 7 4
                    ♣ A K 5
    ♠ J 10 9 2          ┌─────────┐          ♠ 7 6 4
    ♥ 9 7               │    N    │          ♥ A 6
    ♦ K 10 6 3          │ W     E │          ♦ A 8 2
    ♣ Q 10 2            │    S    │          ♣ J 9 8 4 3
                        └─────────┘
                    ♠ A 8 5
                    ♥ K Q J 8 5
                    ♦ Q 9 5
                    ♣ 7 6
```

South	West	North	East
1♥	Pass	4♥	End

When this deal arose in a social game, West led ♠J against Four Hearts and the declarer won with the ace. East won the first round of trumps with the ace and said to himself: 'There's no future in spades or clubs. It looks as if we need three diamond tricks to beat the contract.' He played the ace of diamonds, then another diamond. His partner won with the king but the declarer now had a trick in diamonds, enough to make the contract. Had East left the diamonds alone, the declarer would have had to make the first play in the suit. He would have lost three diamond tricks and gone one down.

'But I thought he might be able to throw a diamond away on the clubs or the spades,' complained East afterwards.

It was not a strong argument. If the declarer could have taken a discard on the clubs or the spades he would have done so before playing on trumps.

Points to remember

1 When you are second to play to a trick, it is usually right to play low.

2 When you are third to play, you should usually play high. An exception is when dummy has an honour and you hold a higher honour. It is then often right to play a lesser card. (For example, you hold ace-jack over dummy's king and play the jack.)

3 Be careful about making the first lead in a new suit. This will often give away a trick.

18 | LEADS, SIGNALS AND DISCARDS

The most important move in defence is the opening lead. If you get off to the wrong start you may never have the chance to recover. Make the right start, and it may be the declarer who is left helpless.

The lead at no-trumps

The opening lead is a particular advantage in no-trumps, because it enables the defenders to develop their own long suit before the declarer can start on his best suit. When you have decided which suit to lead, you must then choose which card to lead. This is the general idea:

- From a suit headed by a sequence (three touching honours) lead the top card: the ace from A K Q, the king from K Q J.
- From a suit headed by a 'broken sequence', such as A K J, K Q 10 or Q J 9, also lead the top card.
- From a suit headed by an 'interior sequence', lead the middle honour: the queen from A Q J, the jack from K J 10.
- Otherwise lead your fourth-best card when the suit contains an honour: the 4 from K J 9 4, the 3 from Q 9 8 3 2.
- When your suit does not contain an honour, lead the second-best card: the 6 from 8 6 3 2, the 8 from 9 8 5 4 2.

When you lead a low spot-card, your partner will expect your suit to be headed by an honour. If instead you lead a high spot-card, he will know that you have chosen a passive lead from a weak suit. This will help him to judge whether he should continue your suit when he gains the lead.

Suppose the opponents' bidding has been 1NT – 3NT and you have to make the opening lead from one of these hands:

(1)   ♠ Q 3   (2)   ♠ J 9 2   (3)   ♠ K J 3
      ♥ K Q J 5 2   ♥ A 4 3   ♥ 9 8 6 2
      ♦ J 7 4   ♦ K J 7 4 3   ♦ A Q 4
      ♣ 10 6 3   ♣ 8 5   ♣ 7 6 3

No problem on (1). You lead ♥ K, the top of a sequence. If instead you
made the mistake of leading ♥ 5, the declarer would score an undeserved
second trick when he held such as ♥ A 10 3. On (2) you lead ♦ 4, the
fourth-best card from a suit containing an honour. On (3) you would lead
♥ 8, the second-best card from a suit not containing an honour. If your
partner gains the lead he will know that your hearts are not very good and
may make a profitable switch to spades or diamonds.

When the opponents have bid some suits

You will not necessarily want to lead your best suit when it has been bid
by the opponents. Suppose this has been the auction:

South	*West*	*North*	*East*
1♥	Pass	2♦	Pass
2NT	Pass	3NT	End

You are on lead with this hand:

                              ♠ J 10 4
                              ♥ K 5
                              ♦ A Q 9 6 2
                              ♣ 9 7 6

There is no point in leading a diamond. The bidding has told you that the
diamonds are strongly held on your left. You must try to find your partner's
long suit. Lead ♠ J.

From a three-card suit you lead the top card when you hold two touching
honours (such as K Q 7 or Q J 5). From three cards to an honour you would
lead the lowest card (the 2 from Q 8 2). From three small cards you would
lead the second-best (the 7 from 9 7 2).

Leading your partner's suit

When your partner has bid a suit, particularly if he has made an overcall,
it is usually right to lead that suit. Indeed his main reason for bidding may
have been to suggest a good lead.

The auction may have been:

South	West	North	East
1♦	Pass	1♥	1♠
1NT	Pass	3NT	End

You are on lead with this hand:

♠ 9 3
♥ J 8 4 2
♦ K 2
♣ Q 10 8 5 2

If your partner had not bid, you would have tried your luck with ♣ 5. Now you should lead ♠ 9 (top of a doubleton) instead. If your partner has a good holding such as K Q J 10 6, your lead will set up the suit.

The lead in a suit contract

There are many more options when you are on lead against a suit contract. You may need to lead your strongest suit, to set up a winner or two there before the declarer can establish some discards. You may instead lead a short suit, hoping that your partner will be able to give you a ruff. On some occasions it will work best to lead a trump, to stop the declarer taking too many ruffs.

As for the choice of card within a suit, it is similar to that against no-trumps. There is one difference, however. Against no-trumps you might lead a low card from a suit headed by two touching honours (A K 7 6 2 or K Q 8 3, for example). Your aim would be to make the long cards later. Against a suit contract you should instead lead the top honour. The third round of the suit will usually be ruffed by one side or another; you must make sure that your honours play a role in the first two rounds.

Attacking in your best suit

When you expect there to be a good side suit in dummy, it is often best to attack in your own strongest suit. This type of hand is very common:

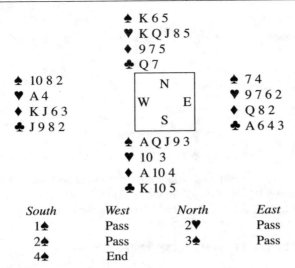

♠ K 6 5
♥ K Q J 8 5
♦ 9 7 5
♣ Q 7

♠ 10 8 2
♥ A 4
♦ K J 6 3
♣ J 9 8 2

♠ 7 4
♥ 9 7 6 2
♦ Q 8 2
♣ A 6 4 3

♠ A Q J 9 3
♥ 10 3
♦ A 10 4
♣ K 10 5

South	West	North	East
1♠	Pass	2♥	Pass
2♠	Pass	3♠	Pass
4♠	End		

West can tell from the bidding that there will be a respectable heart suit in the dummy. Before the declarer can set it up, to obtain discards, the defenders must establish and cash their winners in the other suits.

West's best chance is to attack in diamonds, hoping that his partner will hold the ace or the queen. He leads ♦ 3 and his partner does indeed produce the queen. Declarer will now go down, losing two diamonds and two aces.

If instead West makes a passive lead such as a trump, the declarer will establish dummy's hearts and make the contract easily. Hands such as this are dealt by the million. You will hear players say 'Don't lead from a king.' Nod politely and turn a deaf ear. It is essential to make an attacking lead after this type of auction. Sometimes the declarer will indeed hold ♦ A Q and your diamond lead will be unproductive. But in that case the declarer would very likely have discarded his diamonds on dummy's hearts anyway.

Playing for a ruff

A side-suit singleton is always a promising lead. You will score a ruff when your partner has the ace of that suit, also perhaps if he has a quick entry in trumps.

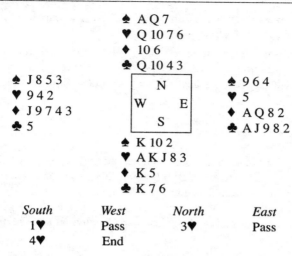

♠ A Q 7
♥ Q 10 7 6
♦ 10 6
♣ Q 10 4 3

♠ J 8 5 3
♥ 9 4 2
♦ J 9 7 4 3
♣ 5

N
W E
S

♠ 9 6 4
♥ 5
♦ A Q 8 2
♣ A J 9 8 2

♠ K 10 2
♥ A K J 8 3
♦ K 5
♣ K 7 6

South	West	North	East
1♥	Pass	3♥	Pass
4♥	End		

Seeing the chance of a ruff or two, West leads his singleton club. East wins with the ace and returns a second club for his partner to ruff. West now wants to give his partner the lead, so he can receive a second ruff. Since the ace of spades is visible in the dummy, West tries his luck with a low diamond. Bulls-eye! East wins with the ace and is now able to lead another club. West ruffs again and the contract is one down.

Trump leads

It is not easy to judge when a trump lead may be best. One situation is when the declarer has been left to play in his second suit:

South	West	North	East
1♥	Pass	1♠	Pass
2♦	End		

If North held, say, two hearts and two diamonds, he would have given preference to hearts (rebidding 2♥). The actual auction suggests that North may hold one heart and three diamonds. In that case the declarer may make extra tricks by ruffing hearts in the dummy. The omens are good for a trump lead.

The same is true on this auction:

South	West	North	East
1♥	Pass	2♥	Pass
2NT	Pass	3♥	End

North has shown a weak hand with ruffing values. You may be able to cut down the declarer's ruffs by leading a trump.

One warning: it is rarely right to lead a singleton trump. You may damage your partner's holding of A J x x, Q J x x or dozens of similar holdings. Left to his own devices, the declarer may misguess the suit and lose an extra trick.

Signals

When your partner leads a high card, such as an ace, you will usually have a choice of spot-cards to play. By playing a high spot-card you can tell your partner that you want the suit to be continued. A low card will suggest instead that you do not want a continuation. This is known as a 'signal'. Suppose your partner leads ♦ A against a spade contract and the diamond suit lies like this:

$$
\begin{array}{c}
\textit{North} \\
♦\,9\,6\,3
\end{array}
$$

West		*East*
♦ A K 10 7		♦ Q 8 2

$$
\begin{array}{c}
\textit{South} \\
♦\,J\,5\,4
\end{array}
$$

Expecting West to hold the ace and king, East will signal encouragement by playing the 8, his highest available spot-card. West will continue with the king of diamonds, then a third diamond.

Suppose instead that the declarer holds the missing queen:

$$
\begin{array}{c}
\textit{North} \\
♦\,9\,6\,3
\end{array}
$$

West		*East*
♦ A K 10 7		♦ J 8 2

$$
\begin{array}{c}
\textit{South} \\
♦\,Q\,5\,4
\end{array}
$$

Here East will signal discouragement with the 2. West will then avoid the mistake of continuing with the king of diamonds, setting up the declarer's queen unnecessarily.

A high-low signal works well also when your partner can ruff the third round.

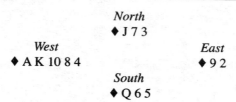

North
♦ J 7 3

West
♦ A K 10 8 4

East
♦ 9 2

South
♦ Q 6 5

West leads ♦ A and East signals encouragement with the 9. West continues with king and another diamond, allowing East to ruff the third round.

There are many other opportunities for the defenders to signal strength in a suit:

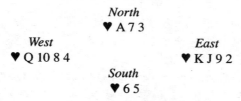

North
♥ A 7 3

West
♥ Q 10 8 4

East
♥ K J 9 2

South
♥ 6 5

West leads ♥ 4 against a suit contract, dummy's ace winning the trick. East signals with the 9 to show that he would welcome a heart continuation, should West gain the lead.

Discards

Another opportunity to pass your partner a message arises when you cannot follow suit. If you throw a high spot-card in another suit, it means that you would like your partner to lead that suit. A low spot-card would mean that you did not want the suit to be led.

The example overleaf will make this clear.

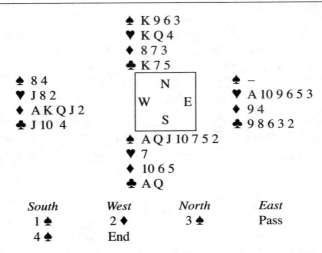

♠ K 9 6 3
♥ K Q 4
♦ 8 7 3
♣ K 7 5

♠ 8 4
♥ J 8 2
♦ A K Q J 2
♣ J 10 4

♠ –
♥ A 10 9 6 5 3
♦ 9 4
♣ 9 8 6 3 2

♠ A Q J 10 7 5 2
♥ 7
♦ 10 6 5
♣ A Q

South	West	North	East
1 ♠	2 ♦	3 ♠	Pass
4 ♠	End		

West starts the defence with three rounds of diamonds. On the third round East has a chance to tell his partner where his strength lies. He throws ♥ 10, a high card to say that he would like West to play a heart. West duly switches to a heart and East scores his ♥ A to put the contract one down. If West had switched instead to a club (or a trump), the declarer would have been able to throw his singleton heart on the third round of clubs.

Points to remember

1 Unless you have an honour sequence, lead fourth-best when your suit contains an honour, or second-best from low cards.

2 Against no-trumps, it is not usually right to lead a suit that has been bid by the opponents. Instead lead a short suit where your partner may have a good holding.

3 Against a suit contract listen to the opponents' bidding and try to guess how the play will go and which type of lead might do the most damage.

4 When your partner leads a suit, you may have the opportunity to signal. A high spot-card will indicate that you like the suit, a low card that you do not like the suit.

5 When you discard on a trick a high spot-card means that you like the suit, a low spot-card means that you do not like the suit.

Quiz

a Against no-trumps, which card would you lead from each of these holdings:

(1) ♠ K J 9 5 4 2
(2) ♦ 9 7 6 2
(3) ♣ K J 10 7 4
(4) ♠ K Q 7 2

b Against a suit contract, which card would you lead from each of these holdings:

(1) ♠ A K 8 7 4
(2) ♥ K 10 7 2
(3) ♦ 8 7 5 3
(4) ♣ Q 2
(5) ♠ K Q 7 2

Answers

a (1) ♠ 5, fourth-best from a suit headed by an honour.

(2) ♦ 7, second-best from a suit with no honour.

(3) ♣ J, middle honour from an interior sequence.

(4) ♠ 2, fourth-best card again.

b (1) ♠ A, top of two touching honours.

(2) ♥ 2, fourth-best from a suit with an honour.

(3) ♦ 7, second-best from a suit with no honour.

(4) ♣ Q, top of a doubleton.

(5) ♠ K. Against a suit contract your honours must play a part in the first two tricks.

19 | DIFFERENT FORMS OF BRIDGE

Until now we have mentioned only the traditional form of the game – rubber bridge. Two other variants deserve a mention: Chicago bridge and duplicate bridge. In this chapter we will look briefly at the proceedings for each type of the game.

Rubber bridge

Unless you have previously agreed who will be partners, you may wish to decide the matter by cutting the cards. The pack is spread face down on the table and each player draws a card. The players drawing the two highest cards will partner each other (using the normal rank of the suits to distinguish between two cards of the same value).

To save time, it is usual to have two packs of cards. The player who drew the highest card may choose a seat (some players are superstitious about such matters) and which pack he will deal. We will say that he chooses to sit South and to deal the blue pack. West shuffles this pack, then passes it to East who will cut the pack into two parts. South places the original lower portion of the pack above the higher portion and deals 13 cards to each player.

When this hand has been played, the deal passes to the left. North, who shuffled the green pack while South was dealing the first hand, will pass the cards to South for the cut. West will restore the cut and deal the second hand. This process continues until the rubber is complete.

Chicago bridge

It is an inconvenience of standard rubber bridge that one rubber may take only two deals (lasting around ten minutes), while another may take many deals (lasting perhaps an hour). Chicago bridge is a variant where a rubber or 'chukker' is of fixed length, consisting of exactly four deals.

The players may cut for partners, as in normal rubber bridge. For the first deal both sides are non-vulnerable. The deal passes to the left at the end of each hand. For the second and third deals, the dealer's side is vulnerable. On the last deal both sides are vulnerable.

When a game is bid and made, that side scores an immediate game bonus: 300 when non-vulnerable, 500 when vulnerable. Part scores may be passed from one deal to the next, unless cancelled by the opponents making a game. A part score made on the last deal of the four, but at no other time, attracts a bonus of 100. If all four players pass on a deal, the cards are re-dealt by the same dealer and the conditions of vulnerability remain the same.

Duplicate bridge (pairs)

To some extent short-term success at rubber bridge depends on whether your side is dealt good cards. In duplicate bridge each deal is played more than once. In a tournament entered by 40 pairs, for example, there will be 20 tables and each deal will be played 20 times. At the end of the event the various results on each deal will be compared. This largely eliminates the luck factor, because you score well only if you achieve better results than other pairs did on exactly the same hands.

The cards are not played in the centre of the table and then gathered into tricks. Instead, each player plays a card by holding it face up in front of him. When the trick is complete, he adds his card, face down, to a line he is building up from left to right. The card will point in the direction of his partner if the trick was won, in the direction of the opponents if the trick was lost. If, at the end of the hand, a player has eight cards pointing in the direction of his partner, and five in the direction of his opponents, he will know that his side scored eight tricks.

Each player then picks up his cards and returns them to a wooden board or plastic wallet. This can then be carried to another table for other competitors to play. Each hand is scored separately. You receive a game bonus of 300 when non-vulnerable, 500 when vulnerable. There is a 50 bonus for a part score.

Examples

1 The board containing the cards for the first deal says that you are non-vulnerable. You make Four Spades with an overtrick. You score +450. That is 150 for the tricks and 300 for a non-vulnerable game. Your opponents on this particular deal will score –450.

2 On the next board, your opponents bid Two Spades and make an overtrick. They score +140 (90 for tricks, 50 for the part score). You will score –140.

The scores are entered on a result sheet held in the board containing the cards. After playing two or three hands you will move to a different table and face different opponents. The eventual scoring for the tournament will be done by the organisers.

Duplicate bridge (teams-of-four)

As well as tournaments for pairs, usually held in a club, you can play in matches between two teams of four players. Each hand will be played only twice. Your team will be North–South at one table, East–West at the other. When the time comes for scoring, you will compare the result you and your partner achieved on each hand with that scored by your opponents on the same cards. If you bid and made Four Spades on one deal, vulnerable, you will have scored +620. If they failed to bid the game and scored only +170, that will be a big gain (or 'swing' as it is called) for your team.

The difference in scores on each hand is converted into International Matchpoints (or IMPs), according to a table. It would be inappropriate to give details here but the difference of 450 in the example we just considered would be worth ten IMPs. You might then say, 'We gained ten IMPs by bidding the spade game on Board 5.' At the end of the match both teams add their IMPs gained and lost to see who has won. This is the form of the game used for all major championships, including the Olympiad and the Bermuda Bowl.

GLOSSARY OF BRIDGE TERMS AND PHRASES

(The glossary is confined to terms mentioned in this book.)

Above the line   All scores other than those for tricks bid and made belong above the line drawn across the centre of the scoresheet.

Balanced   A balanced hand is one containing no singleton or void, usually 4–3–3–3, 4–4–3–2 or 5–3–3–2.

Below the line   Scores for tricks bid and made, such as 90 for Three Hearts bid and made, are entered below the centre line on the scoresheet.

Bid   An undertaking to take a specific number of tricks, either with a chosen trump suit or at no-trumps. For example, Two Spades means you think you can made eight tricks with spades as trumps.

Blackwood   A conventional bid of 4NT after which your partner must reveal how many aces he holds.

Call   A term covering any bid, pass, double or redouble.

Clear   To clear a suit is to drive out all the winners held by the opponents.

Combination finesse   A finesse where two adjacent cards are missing, such as in A J 10.

Communication   The ability to go from hand to hand.

Contract   The final call determines the contract in which the hand is played – for example, Four Spades doubled.

Control   A holding that will prevent the opponents from scoring tricks in a suit (ace, king, singleton or void).

Convention   An agreement between partners to use a bid in a sense not obvious on the surface.

Cue bid   (a) A bid in an opponent's suit, usually to show strength. (b) A bid which shows a control, such as an ace, rather than a suit.

Cut   Divide the pack, prior to the deal.

Cut for partners   The moment when the four players each draw a card to see who will be partners.

Declarer   The player who must attempt to make the contract, playing the dummy's cards as well as his own.

Defenders   The two players who attempt to stop the declarer from making his contract.

Denomination   The trump suit, or no-trumps.

Discard   The play of a card (not a trump) which does not belong to the suit led.

Distribution   The pattern of suit lengths in a player's hand (for example, 5–4–3–1).

Double   A call that increases the penalties if a contract is not made, also the bonuses if it is made.

Double finesse   A finesse that seeks to entrap two cards, as when you lead to the 10 from A Q 10.

Double raise   A raise that covers two steps – when you raise One Heart to Three Hearts, for example.

Doubleton   A holding of two cards in a suit.

Drop   To cause an opposing high card to fall by playing higher cards.

Dummy   (a) The partner of the declarer. (b) The hand exposed opposite the declarer.

Duck   To play low when you hold a higher card.

Entry   A high card used to cross from one hand to the other.

Establish   To make a suit good by removing the opponents' high cards.

Finesse   An attempt to win a trick with a lesser card in a tenace. You hope that the outstanding higher card lies to the left.

Forcing bid   A bid that requires your partner to bid again.

Forcing to game   A bid that requires both partners to continue bidding until game is reached.

Game   To make a game, you must score 100 points below the line.

Grand slam   A contract to win all 13 tricks.

Guard   A high card representing a 'stop' in one of the opponents' suits.

Hold-up   To refuse to part with a high card.

Honour card   Ace, king, queen, jack or 10.

Intervening bid   A bid by the side that did not open the bidding.

Jump   A bid, rebid, raise or response made one level higher than necessary.

Knock out   Play on a suit with the aim of removing a defender's high card.

Lead   To play the first card to a trick.

Limit bid   A bid which defines the strength of your hand within narrow limits.

Major suit   Spades or hearts. Four of a major produces game.
Minor suit   Diamonds or clubs. Five of a minor produces game.
No bid   Call that denotes a pass. (In the USA the word 'Pass' is used).
No-trumps   Denomination in which there is no trump suit.
Open the bidding   To make the first bid (not a pass).
Overbid   A bid that overstates the value of the hand.
Overcall   The first bid made by the side that did not open the bidding.
Overruff   To play a higher trump than that of a player who has already ruffed.
Overtrick   A trick in excess of the contract.
Part score   A contract scoring less than 100 below the line.
Pass   Call indicating that the player does not want to bid, double or redouble.
Penalty   Points scored above the line when the opponents' contract has failed.
Penalty double   A double seeking to increase the penalty when the opponents' contract fails.
Point count   A method of valuation in which points are assigned to aces, kings, queens and jacks.
Pre-empt   To make a high call on a weak hand with a long suit. The aim is to prevent the opponents from bidding accurately.
Preference   A bid or pass which indicates which of your partner's bid suits you prefer.
Raise   To make a higher bid in a suit just bid by your partner.
Rebid   (a) The second bid made by a player. (b) To bid again a suit that you have already bid.
Redouble   Either member of a side that has been doubled may redouble. This increases the scores if the contract is made, also the penalties if the contract fails.
Responder   (a) The partner of the opening bidder. (b) The player who responds to any specific call.
Reverse   A player reverses when he bids at the two level a suit higher-ranking than his first suit (as in 1♦ – 1♠ – 2♥).
Rubber   A rubber is won by the side that first scores two games.
Ruff   To play a trump when a side suit has been led.
Sacrifice   To bid a contract which you expect to fail, with the aim of conceding a penalty less than the value of the opponents' contract.
Sequence   A group of consecutive cards, usually honours, such as K Q J.

Show out   To fail to follow, having none of the suit led.

Side suit   A suit other than the trump suit.

Single raise   Usually a raise from One to Two.

Singleton   The holding of only one card in a suit.

Slam   A contract to make 12 or 13 tricks.

Small slam   A contract to make 12 tricks.

Stayman   A conventional bid of 2♣ over your partner's 1NT, asking him to bid a four-card major suit.

Stopper   A high card that will prevent the opponents from running a suit (usually at no-trumps).

Support   (a) To raise your partner's suit. (b) Trump support is your holding in a suit bid by your partner.

Take-out double   A double that asks your partner to choose a trump suit (or bid no-trumps), also to indicate how strong his hand is.

Tenace   Two non-touching high cards, such as A Q or K J.

Trick   A trick consists of four cards, one in turn from each player.

Trump   A suit, determined in the bidding, which has the power to beat any card in a different suit.

Underbid   A bid that understates the value of the player's hand.

Undertrick   The trick or tricks by which the declarer fails to make his contract.

Void   A holding of no cards in a suit.

Vulnerable   A side that has won a game becomes vulnerable. Penalties and some bonuses are then increased.

USEFUL ADDRESSES

To obtain more information about bridge or to inquire about clubs in your area, contact your national association.

American Contract Bridge League
2990 Airways Boulevard
Memphis TN 38116–3847
USA

Australian Bridge Federation
PO Box 3222
Manuka ACT 2603
Australia

Canada: as for USA above.

English Bridge Union
Broadfields
Bicester Road
Aylesbury
Bucks
HP19 3BG
England

New Zealand Bridge/
Contract Bridge Association
PO Box 12116
Wellington New Zealand

Scottish Bridge Union
32 Whitehaugh Drive
Paisley
PA1 3PG
Scotland

South African Bridge Federation
PO Box 87682
Houghton
Johannesburg
South Africa

INDEX

ty TEACH YOURSELF

CARD GAMES

David Parlett

The delightful thing about cards is their variety and vitality. There are literally hundreds of different card games available to appeal to any and all tastes – from the intellectual intricacies of bridge to the psychological excitement of poker.

This book includes classic games from the past, such as piquet and bezique; national games from abroad, such as skat and preference; timeless favourites like rummy; and modern inventions and discoveries, such as ninety-nine and barbu. It has been specially compiled for three types of players:

- for beginners, who have never touched a card before but are willing to find out what fun they have been missing;
- for regulars, who play one or two favourites regularly and need a reference to resolve disputed points;
- for explorers, who enjoy experimenting with card games from different times and places.

For all those prospective players David Parlett is the ideal teacher and guide. A free-lance games writer and consultant, he has been collecting, inventing, and writing about card games for most of his life.

ty TEACH YOURSELF

CARD GAMES FOR ONE

David Parlett

Patience is to mental exercise as jogging is to physical, being a most effective way of toning up the brain and abolishing mental flab.

Card expert David Parlett has been collecting, comparing, arranging and inventing Patience games for over 20 years, and here describes his favourites with lucidity and wit.

But which is the game for you? Do you even know how large the choice is? *Teach Yourself Card Games for One* offers a varied selection of 150 or so of the choicest games from a seemingly unlimited treasury of human ingenuity.

They include:

- games for one pack, two packs and stripped packs;
- games of skill to test your intellect and games of chance that try your patience;
- games that seek to please by always 'coming out' and games that do their best to get you down;
- games ranging in design from opulent Victorian extravaganzas to post-modern minimalist 'abstracts'.

TEACH YOURSELF

CHESS

Bill Hartston

This revised and updated book contains all you need to know to learn and develop an understanding of good chess. The early chapters describe the moves of the pieces, elementary tactics in attacking and defensive play, and the basic combinations to force checkmate. The book then goes on to explain how to fight for control of the board in the opening stages, how to assess and exploit positional strengths and weaknesses and how to develop a strategic plan, leading from the opening into middlegame and endgame. After analysing the most common opening variations, Bill Hartston provides detailed commentaries on a series of historic games, selected to illustrate different styles and strategies of play. This new, enlarged edition includes added exercises for the reader, and is brought right up to date with a discussion of the battle for chess supremacy between humans and computers.

Whether you are a complete beginner or seeking to improve your present level of play, this book will deepen your understanding of chess and enhance your enjoyment of the game, as player or spectator.

elated titles

BACKGAMMON

Robin Clay

Backgammon is compulsive, easy to learn and quick to play. It is a game which offers hours of entertainment with an exciting blend of chance and skill.

This book explains the basics of backgammon concisely and clearly, and then goes on to discuss tactics and aggressive and defensive styles of play. Every game of backgammon is unique, so Robin Clay encourages you to experiment, take risks and try out new strategies. The social version of the game – known as Chovette – which can be played by up to six players on one backgammon board, is also fully explained.

Whether you play it as a simple game of luck or a battle of skill and deception, backgammon combines the intellectual stimulation of chess, bridge and Go with the excitement of knowing that your fate depends on the roll of the dice.